Wolfgang Appelt

Document Architecture in Open Systems: The ODA Standard

With 77 Figures

Springer-Verlag

Berlin Heidelberg New York
London Paris Tokyo
Hong Kong Barcelona
Budapest

Dr. rer. nat. Wolfgang Appelt
Institute for Applied Information Technology
National Research Center for Computer Science (GMD)
Schloss Birlinghoven
W-5205 Sankt Augustin 1, FRG

ISBN-13:978-3-642-76922-1   e-ISBN-13:978-3-642-76920-7
DOI: 10.1007/978-3-642-76920-7

© Springer-Verlag Berlin Heidelberg 1991
Softcover reprint of the hardcover 1st edition 1991
Typesetting: camera-ready by author.

45/3140-543210 – Printed on acid-free paper

Preface

In 1989, the International Standard *"Office Document Architecture (ODA) and Interchange Format"* was published by ISO (*International Organization for Standardization*) with the registration number ISO 8613. Also, a *Recommendation* of the CCITT (*Comité Consultatif International Télégraphique et Téléphonique*) with essentially the same technical content was issued in 1988 within the T.410 series with the title *"Open Document Architecture (ODA) and Interchange Format"*.

These standards were developed for the electronic interchange of documents which are stored in digital form. The documents envisaged are primarily, but not exclusively, those which are encountered in an office environment such as letters, memos, reports, contracts or invoices. The receiver of such a document should be able to render the document on a printer or a computer screen in the form specified by the sender, but he or she should also be able to process the document, for example, to modify its content or its visual appearance.

The origins of ISO 8613 reach back to the early 1980s when the development of standards for electronic document interchange started. The first efforts leading to ISO 8613 in its present form started 1982 within ECMA, the *European Computer Manufacturers Association*, which published the ECMA Standard 101 in 1985. ECMA 101 contained already all the essential concepts of ISO 8613. (In the meantime, ECMA 101 has been aligned to ISO 8613.)

The ECMA project initiated parallel projects within ISO (assigned to *Working Groups* 3 and 5 of TC 97/SC 18 which was established in 1981) and CCITT (assigned to *Study Group* VIII which formed a close liaison with SC 18 in 1983). In 1984, CCITT published the Recommendation T.73 which can also be considered a predecessor of ISO 8613 and the T.410 series of Recommendations. Within the ISO Reference Model for *Open Systems Interconnection* (OSI) ISO 8613 belongs in layer 7 (*Application Layer*).

In 1990, the name of ISO 8613 was changed to *"Open Document Architecture (ODA) and Interchange Format"*, firstly, to align with the

CCITT Recommendations and, secondly, to express the broader scope of the Standard which is not only directed to the office world, though the origins of the Standard come out of this area. ISO/IEC JTC 1/SC 18, the successor of ISO TC 97/SC 18, decided that this name change will officially take place with the republication of ISO 8613, probably in 1992. At present, a reference to ISO 8613 both with the terms *"Office"* and *"Open"* can be considered correct.

This book gives an introduction to the ISO Standard 8613, but it can also be regarded as an introduction to the CCITT Recommendations of the T.410 series since both are technically almost identical. It is primarily addressed to readers who want to explore the applicability of the Standard to their needs, who plan to introduce systems conforming to the Standard within their document processing environment or who intend to develop systems based on the Standard. In the latter case, a thorough reading of the Standard itself is necessary, of course: this introduction cannot deal with all details of the Standard since ISO 8613 comprises about 700 pages at present.

Furthermore, it should be noted that extensions of ISO 8613 are currently under development and also modifications to the existing text are to be expected, for instance, when errors in the Standard are found. This book describes the state of the Standard in Spring 1991, reflecting the extensions and modifications since the first publication of the Standard in 1989.

The structure of the book follows essentially the structure of the Standard itself which is divided into eight parts. Between these parts there exist sometimes rather complex interrelations. When reading this book for the first time, it is therefore not sensible to read the book completely from the first to the last page, since sometimes details will only be understandable when terms and concepts introduced at later places in the book are already known. For a first reading, it is therefore recommended to read Chap. 1 ("Structure of the Standard") and from the other chapters only the first section (i.e., 2.1, 3.1, etc.). Afterwards the reader should have a sufficient overview of the whole Standard to allow a complete reading of the book.

Sankt Augustin, May 1991   Wolfgang Appelt

Table of Contents

1 The Structure of the Standard

The ODA Standard – the widely known acronym ODA will often be used when referring to ISO 8613 – consists of eight parts:

Part 1: *Introduction and general principles*
Part 2: *Document structures*
Part 4: *Document profile*
Part 5: *Office document interchange format (ODIF)*
Part 6: *Character content architectures*
Part 7: *Raster graphics content architectures*
Part 8: *Geometric graphics content architectures*
Part 10: *Formal specifications*

At present, Parts 3 and 9 do not exist. During the development of ISO 8613 the Standard was structured a little differently, including also a Part 3. At a rather late stage, however, when Parts 6, 7 and 8 were almost finished, a restructuring was considered necessary, making the previous Part 3 obsolete. Given the time constraints a renumbering of the parts was not carried out since, due to the numerous cross-references between the parts, this would have required major editorial modifications. (There is also no CCITT Recommendation T.413.)

Part 10 of the Standard was published 1990, about one year after the first publication of ISO 8613. The formal specifications did not become Part 9 since this number was already assigned to a further content architecture, namely audio content, i.e., it is expected that Part 9, *Audio content architectures*, will be published. Work on this Part 9 has already started.

Between these parts of the Standard there are more or less extensive cross-relations. The first part, *Introduction and general principles*, provides a general overview of the Standard and contains especially numerous definitions of terms used in this and other parts of the Standard.

Part 2, *Document structures*, can be regarded as the central part of the whole Standard in which the general structural concepts for ODA documents are specified. In this part, however, the actual content of documents is not described in detail. This is done in Parts 6, 7 and 8, where the processing for three kinds of information within ODA documents, namely of "text" in its proper meaning, of raster graphics and of geometric graphics is specified. (The term "text" has in ISO 8613 usually a broader meaning, including also graphical information.)

Part 2 is by far the most voluminous of the Standard and therefore in this book the chapter dealing with Part 2 is rather long compared to the others since understanding the *document structures* is the key for understanding the whole Standard.

This structure of ISO 8613, especially the separation between the structural concepts of ODA documents in Part 2 and the different kinds of content in Parts 6, 7 and 8, provides for the extension of the Standard by further content types, for instance digitized voice, without requiring major modifications of already existing parts. At least, this is the main rationale for this structure.

The main purpose of the *Document profile* described in Part 4 is to store, and maybe process, a certain number of pieces of information describing the global properties of an ODA document separately from the document itself.

Part 5, *Office document interchange format (ODIF)*, specifies in which way ODA documents have to be encoded for the interchange with other partners, i.e., Part 5 specifies the data format which ODA systems must be able to accept and generate. Certain parts of ODIF concerning the encoding of text, raster graphics and geometric graphics, however, are specified in Parts 6, 7 and 8.

Following the regulations for ISO Standards, all parts of ISO 8613 have a similar structure. After a table of contents and a foreword, the scope of the Standard as a whole and of the particular part are given. The section *Normative references* lists those documents which are referenced and thus become implicitly also a part of ISO 8613. Only ISO Standards and CCITT Recommendations are referenced.

Afterwards the actual technical specifications follow. Part 1 starts here with a comprehensive list of definitions for terms used in the Standard. According to the ISO regulations the other parts also contain a section *Definitions* but it usually consists only of a reference to the definitions in Part 1.

All parts of the Standard contain additional annexes. There is a distinction between *normative annexes* which have the same normative quality as the specifications in the body parts and whose content has to be considered when implementing the Standard, and between *informative annexes* which are not part of the technical specifications. The latter annexes often contain explanatory material which is provided for a better understanding of the specifications.

As mentioned above, several ISO Standards and CCITT Recommendations are implicitly part of ISO 8613 and are referred to as normative references. The can be classified as follows:

ISO Standards specifying the coding of character sets:

ISO 646:   *Information processing – ISO 7-bit coded character set for information interchange* (1983)

ISO 2022:   *Information processing – ISO 7-bit and 8-bit coded character sets – Code extension techniques* (1986)

ISO 2375:   *Data processing – Procedures for registration of escape sequences* (1985)

ISO 6429:   *Information processing – ISO 7-bit and 8-bit coded character sets – Additional control functions for character-imaging devices* (1983)

ISO 6937:   *Information processing – Coded character sets for text communication – Part 1: General introduction* (1983)

Information processing – Coded character sets for text communication – Part 2: Latin alphabetic and non-alphabetic graphic characters (1983)

ISO 7350:   *Text communication – Registration of graphic character sub-repertoires* (1984)

An ISO Standard specifying the representation of dates and times:

ISO 8601:   *Data elements and interchange formats – Information interchange – Representation of dates and times* (1988)

An ISO Standard specifying a data format for graphical information:

ISO 8632:   *Information processing systems – Computer graphics – Metafile for the storage and transfer of picture description information – Part 1: Functional specification* (1987)

Information processing systems – Computer graphics – Metafile for the storage and transfer of picture description information – Part 3: Binary encoding (1987)

ISO Standards specifying a syntax for the coding of data structures:

ISO 8824:   *Information processing systems – Open Systems Interconnection – Specification of Abstract Syntax Notation One (ASN.1)* (1987)

ISO 8825:   *Information processing systems – Open Systems Interconnection – Specification of basic encoding rules for Abstract Syntax Notation One (ASN.1)* (1987)
Information processing systems – Open Systems Interconnection – Specification of basic encoding rules for Abstract Syntax Notation One (ASN.1). Addendum 1: ASN.1 extensions

Additional ISO Standards for document processing:

ISO 8879:   *Information processing – Text and office systems – Standard Generalized Markup Language (SGML)* (1986)
ISO 9069:   *Information processing – SGML support facilities – SGML Document Interchange Format (SDIF)* (1988)

An ISO Standard for character fonts:

ISO 9541:   *Information processing – Font and character information interchange – Part 1: Architecture*
Information processing – Font and character information interchange – Part 2: Interchange format

CCITT Recommendations for facsimile documents:

CCITT Recommendation T.4:   *Standardization of Group 3 facsimile apparatus for document transmission* (1988)
CCITT Recommendation T.6:   *Facsimile coding schemes and coding control functions for Group 4 facsimile apparatus* (1988)

Several of these Standards are not referenced completely but only in part as can be seen in the list above. Furthermore, some of the Standards were not yet published when ISO 8613 appeared; these are the Standards for which no publication year is given in the list above.

2 Part 1: Introduction and General Principles

This part of the ODA Standard has essentially the following technical content:

- It contains a detailed list of definitions of the terms used in the standard.
- It gives an introduction to the "world" of ODA, especially to the basic concepts of ODA documents.
- The contents of the other parts of the Standard are briefly described and cross-relations between the parts are pointed out.
- The concept of the *document application profiles* which is referenced from other parts of the Standard is explained.
- Conformance specifications for ODA documents are defined.

In this chapter the basic concepts and the document application profiles will be explained in greater detail.

2.1 The Basic Concepts of ODA

The ODA Standard is based on several practical assumptions about what documents (in the world of ODA) are and which structures are present in such documents. These ideas will be discussed in this section. As a first step, only a rather coarse overview will be given here with more details on the concepts following later, especially in Chap. 3.

2.1.1 Logical Structures and Layout Structures

The basic proposition of ODA is the existence of a logical view and a layout view of a document.

If a document is looked at from a layout viewpoint the observer is interested in the physical appearance of a document on a rendition medium, for example on paper or on a computer screen. In this case the *layout structure* of a document is identified. For instance, it may be perceived that the document consists of a number of pages, that the pages share a similar graphical make-up, e.g., they might contain headlines and footlines and the actual text might be set in three columns. The objects identified by this layout view are called *layout objects*.

If a document is looked at from a logical viewpoint the observer is interested in the *logical structure* into which the content of the document is divided. For instance, it may be perceived that the document consists of several chapters, each chapter being substructured into sections, each section being a sequence of paragraphs or pictures. The objects identified by this logical view are called *logical objects*.

These two alternative, but complementary structures in documents, the logical structure and the layout structure, are the basis for ODA's so-called *document architecture*. ODA does not permit any arbitrary ordering for the the elements of the structures (the logical objects and the layout objects) but requires that the elements within the logical structure and within the layout structure build a so-called *tree structure.*

The root node of the logical structure is called the *document logical root*, the root node of the layout structure is called the *document layout root*. The nodes of the trees represent the objects within a document. The terminal nodes of the trees, i.e., the nodes having no sub-nodes, are called *basic objects*, or *basic logical objects* and *basic layout objects* if the structure to which an object belongs is explicitly mentioned. Correspondingly, the other nodes of the trees are called *composite objects*, *composite logical objects* or *composite layout objects*, respectively.

Considering the logical structure of a document, all kinds of objects thus have been introduced: an ODA document contains exactly one document logical root, one or more composite logical objects and one or more basic logical objects.

Considering the layout structure of a document, there are three more kinds of objects besides the document layout root, namely *blocks, frames, pages* and *page sets*.

A *block* is a rectangular area on the rendition medium. It contains a part of the actual content (text or graphics) of the document and is always a basic layout object, i.e., it has no subordinate objects.

A *frame* is also a rectangular area on the rendition medium. It is subdivided into either one or more frames or one or more blocks. A mixture of frames or blocks is not allowed. A frame is always a composite layout object since it has subordinate objects.

A *page* is also a rectangular area on the rendition medium and corresponds essentially to the common understanding of this term, i.e., a sheet of paper or the simulation of a sheet of paper on a computer screen. A page is usually a composite layout object, having either one or more frames or one or more blocks as subordinate objects. A mixture of frames and blocks is not allowed. Sometimes a page can also be a basic layout object, i.e., it may have no subordinate objects, as will be seen later.

A *page set* is a set whose elements are pages or page sets. A mixture of pages and page sets is allowed. A page set is therefore also a composite layout object with pages and page sets as subordinate objects.

The *document layout root* is at the top level of the hierarchy in the layout structure and has page sets or pages or a mixture of both as subordinate objects, i.e., it is also a composite layout object.

For a better understanding of these terms consider Fig. 1 which shows a business letter. If this letter is looked at from a layout viewpoint one might reach the following conclusions using ODA terminology (see Fig. 2):

The letter is a page containing three frames (the header area, body area and end area). Each of these frames contains several blocks which have no further substructure. The actual content of the letter is associated with these blocks, e.g., the company logo is associated with the logo area and the signature of the sender with the signature area.

It should be noted that this layout structure of the document is just one example and does not necessarily follow from the appearance of the document. From ODA's point of view it would also be correct to omit the substructuring of the page into the three frames and regard the twelve blocks as direct subordinates of the page. (However, it would not be correct to keep the header area and text area as frames but omit the frame containing the end area and have these blocks as direct subordinates of the page, since a page cannot have blocks and frames as subordinates simultaneously.)

As mentioned above, the layout structure of the letter builds a tree structure as shown in the upper half of Fig. 3. The document layout root for this example is "letter layout" whose subordinate is a page called "letter area". This page has the three subordinate frames (composite layout objects) "header area", "body area" and "end area" which in turn have the subordinate blocks (basic layout objects) "address area" ... "logo area", "salutation area" ... "3rd paragraph area" and "greetings area" ... "enclosure area". To each of the basic layout objects belongs a certain piece of the actual content of the letter. These pieces of content are called *content portions* in ODA terminology and are represented by the crossed-out circles in Fig. 3.

Mr.
R. Eader
12, Tree Road

Ipswich   24 April 1990

Subject: Structures in ODA documents

Dear Mr. R. Eader,

This small example shall show some of the structures
in ODA documents.

This is the second paragraph of the body text of
this letter.

And this is the last paragraph before the end of
the letter.

                                        Best Regards

                                        A. UTHOR

                                        (A. Uthor)

Enclosure: ISO 8613

Fig. 1: Example of a business letter

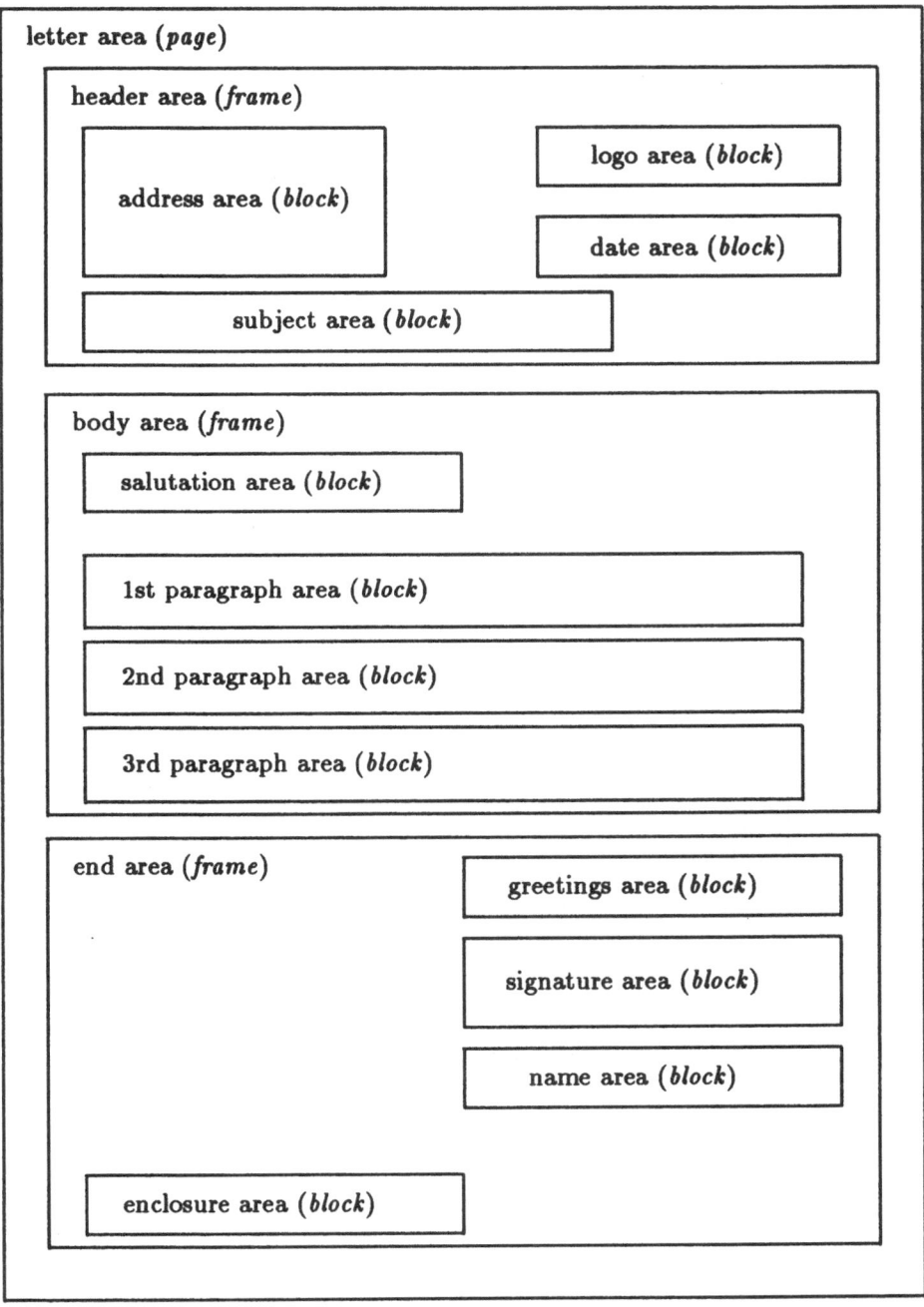

Fig. 2: Layout structure of the business letter

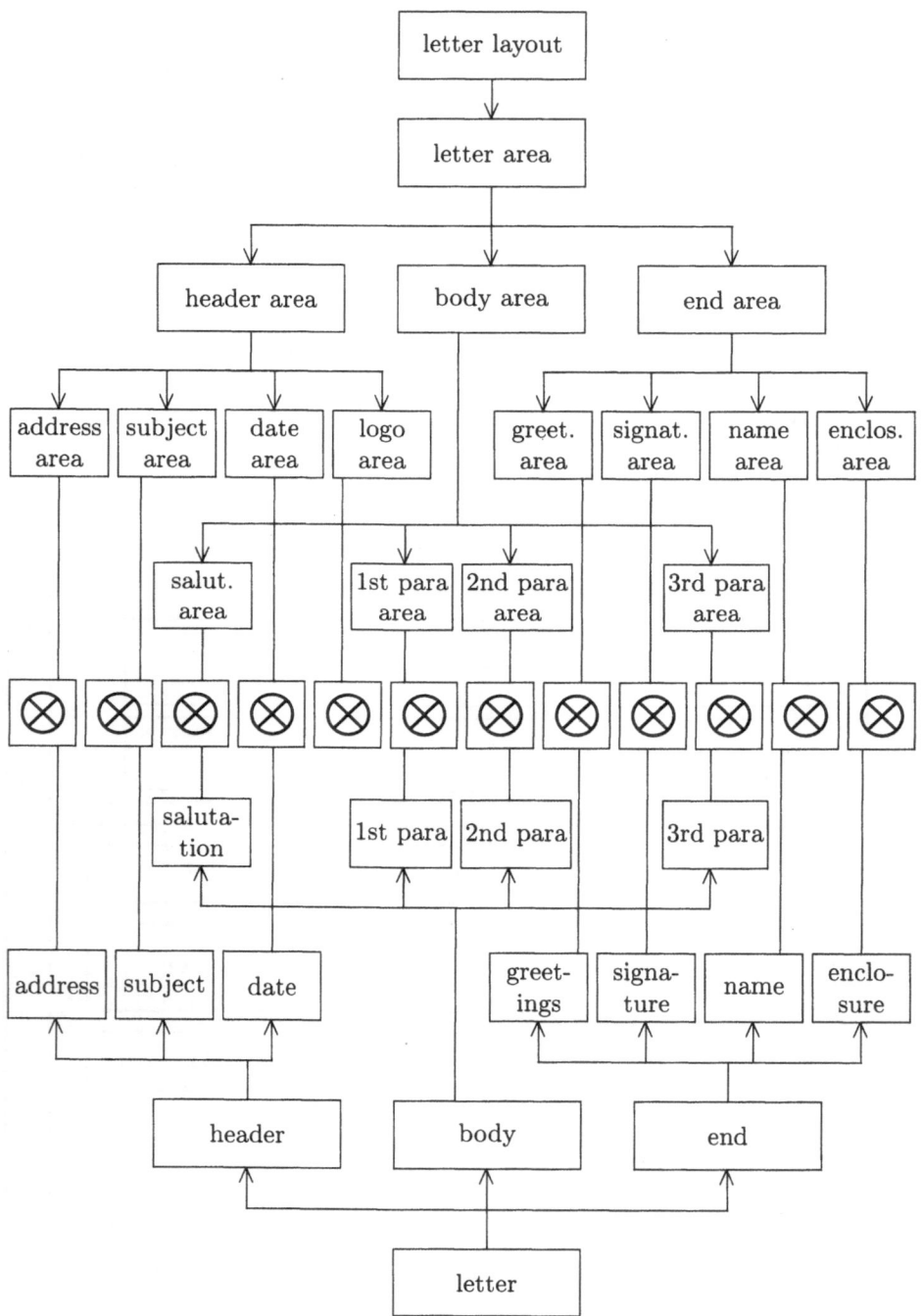

Fig. 3: Tree structure of the business letter

Before continuing the discussion of Fig. 3, the following should be noted: The names of the layout objects such as "letter area", "address area" or "logo area" in this example suggest a certain semantics for these objects, however, this semantics is not present from ODA's point of view. Regarding ODA's document structure these objects are just a document layout root, composite layout objects or basic layout objects. Specifying semantics to objects is outside the scope of the ODA Standard and can only be done by additional agreements between "users" of ODA documents by means which are not defined in the Standard.

The lower half of Fig. 3 shows the logical structure of the letter which builds also a tree structure. A "letter" (document logical root) consists of the three composite logical objects "header", "body" and "end" which in turn consist of the basic logical objects "address" ... "date", "salutation" ... "3rd paragraph" and "greeting" ... "enclosure". Again, to each of the basic logical objects belongs a certain piece of content (content portion) of the letter.

Similar comments as made for the layout structure apply also to the logical structure: Firstly, this logical structure for the letter is just one example; a reader of the document might perceive a slightly different logical structure, for instance, he might regard the "salutation" as logically belonging to the header of the letter. Secondly, though the names of the objects suggest a certain semantics for them, these semantics are not present from ODA's point of view; ODA defines only a document layout root, composite logical objects and basic logical objects as components of the logical structure. The semantics of these objects in a certain document is outside the scope of the Standard.

A closer look at Fig. 3 reveals some further properties of ODA documents. It can be observed that a content portion is associated with each basic object, both in the logical structure and the layout structure. This is not an accidental property of this example but a general architectural principle of ODA documents: With each basic object a certain piece of the actual content of the document is associated.

Furthermore, it can be seen that the content portions are a kind of interface between the logical structure and layout structure. Again, this is a basic principle of ODA's document model for obvious reasons: Every piece of content of a document which cannot be subdivided further and which belongs to a basic logical object must also be associated with a basic layout object to be visible on the rendition medium.

On the other hand, an ODA document may have content portions which are only associated with a basic layout object, not with a basic logical object. In the example shown, this is true for the company logo which belongs to the basic layout object "logo area". This content portion is not associated with a basic logical object. Business letters, for instance, are often written on form sheets that already have a certain content, e.g., the company logo is usually already printed on these sheets. The author of a business letter will usually not consider the predefined content on the form sheets as part of the logical structure of his letter. A logo, for instance, is for this reason not part of the logical structure of the document; however, it is part of the layout structure since it appears on the printed sheet.

It should be noted that the logical structure and the layout structure of a document can be quite different. The great similarity between these two structures in the example above is mainly due to the fact that the document in the example has a rather simple structure.

Until now it was implicitly assumed that a document always has both a logical and a layout structure. In principle, this is true, of course, since a document can only be read by a human if it is displayed on a rendition medium, e.g., on paper or on a computer screen. It therefore has a layout structure. Conversely, any document – if it is not just an arbitrary collection of letters without any meaning – has some intellectual content and therefore some logical structure.

When talking about ODA documents, however, it is always assumed that these documents are stored in electronic form, and such documents may not have a logical structure or a layout structure. This means the following: In ODA terminology, a document has no layout structure if the stored document contains no information on how the document is to be represented on a rendition medium, and if only the actual content of the document and its logical structure are available in electronic form. Before such a document can be perceived by a human reader, a layout structure for the document must be supplied so it can be displayed on a rendition medium. However, it can be sensible to use a document without a layout structure in electronic interchange. The receiver gets all the information that represents the intellectual content of the document, although before he or she can read it he or she must supply an appropriate layout structure.

Correspondingly, an ODA document has no logical structure if the stored document contains no information on how the content of the document is logically structured, and if only the information describing its physical appearance on a rendition medium is available in electronic form. As an extreme example, one might consider a document whose pages are stored in facsimile format, i.e., essentially in the form of raster images. It can be sensible to use such a document in electronic interchange, as the widespread use of fax transmission in the business world shows. The receiver can display such an ODA document on a rendition medium and, by reading it, perceive its intellectual content. He or she will then usually also recognize logical structures in the document; however, it is unlikely that these will be the same as those the author of the document had in mind during its preparation.

It should be noted that some structural aspects of ODA documents have been described so far, but nothing has been said about how these structures are represented in electronic form. This will be explained in detail in Sect. 3.1.

We conclude this section with some remarks on the content portions which were mentioned at several places above. The content portions contain the actual content of ODA documents. All the other objects introduced so far contain essentially only information describing the structuring of the elements of a document.

The internal structure of the content portions is described in Parts 6 (character content architectures), 7 (raster graphics content architectures) and 8 (geometric graphics content architectures) of the Standard. Each content portion has exactly one content architecture; i.e., a mixture of text and raster graphics, for instance, within one content portion is not possible. However, several content portions can be associated with one logical object, though this is not shown in the business letter example above. As a first approach to ODA's architectural model, the content portions can be regarded as rectangular areas (from a layout point of view) or as pieces of content elements of a document which have no further substructure (from a logical point of view), each content portion containing a piece of text, a raster image or a geometric image.

In the business letter, for instance, the block called "logo area" might be associated with a content portion containing a geometric image and the block called "signature area" with a content portion containing raster graphics, assuming that the signature was digitized by a scanner. The other content portions contain text (character content).

2.1.2   Document Classes

It is often the case that certain documents have similar structural properties. For example, a letter usually has an address, a date, a salutation and a signature, i.e., there are some common rules for the logical structure of letters. Also the physical appearance of letters often has some common properties. For instance, the places where the address, the date or the salutation appear on the sheet are usually fixed.

To handle common structural properties of documents the concept of *document classes* has been introduced into the ODA Standard. A document class is the specification of a set of rules concerning the logical structure and the layout structure of documents belonging to this document class. In other words: A document belongs to a given document class if its logical structure and layout structure conform to the specifications given in the document class.

The rules which are given in a document class for the logical structure are called the *generic logical structure*, and the rules which are given in a document class for the layout structure are called the *generic layout structure*. Correspondingly, the structures in a specific document are called the *specific logical structure* and *specific layout structure*. In Sect. 2.1.1 therefore, the adjective "specific" should have been added to these terms, but this was deliberately not done there to make the introduction to the basic concepts of Sect. 2.1.1 easier.

All objects of the specific structures have corresponding object classes in the generic structures. The generic logical structure of a document class contains *logical object classes* which are either a *document logical root class, composite logical object classes* or *basic logical object classes*. The generic layout structure of a document class contains *layout object classes* which are either a *document layout root class, page set classes, page classes, frame classes* or *block classes*.

For a better understanding of the ODA concept of a document class, consider Fig. 4 which shows the generic logical structure of a business letter.

The graphical notation used in the upper part of the figure for describing the generic logical structure is explained in the lower part. According to the notation the figure is to be interpreted as follows: A business letter conforming to the rules of this generic logical structure consists of three composite logical objects called "header", "body" and "end". The "header" consists of the basic logical objects "address", "date" and, optionally, "subject". The "body" consists of one or more composite logical objects called "body content" each of which is either a basic logical object called "text" or a composite logical object called "figure". A "figure" consists of the two basic logical objects "picture" and "label". The "end" consists of the basic logical objects "greetings", "signature", "name" and, optionally, "enclosure".

It should be noted that the specific logical structure of Fig. 3 does not conform to this generic logical structure: In the specific logical structure of the business letter the subordinate objects of "body" are basic logical objects whereas the subordinate objects of "body" in this generic logical structure are composite logical objects. Therefore, the specific document

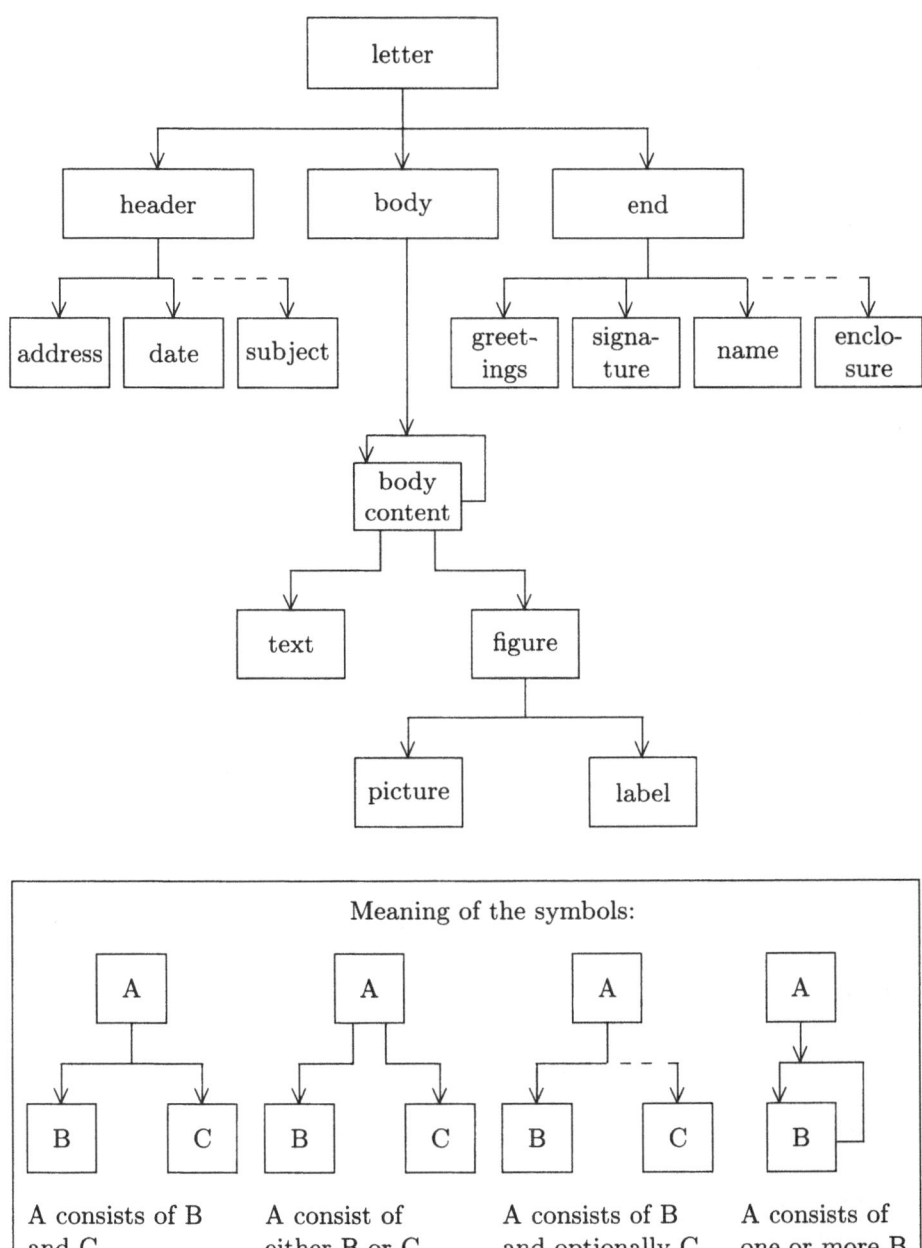

Fig. 4: Generic logical structure of a business letter

whose structures are shown in Fig. 3 does not belong to the document class whose generic logical structure is shown in Fig. 4.

The specific logical structure and the specific layout structure of ODA documents are always tree structures (see Fig. 3) but this is not necessarily the case for the generic structures as can be seen in Fig. 4. However, the generic logical structure as well as the generic layout structure must be built in such a way that the specific structures which can be derived from them are tree structures. The precise construction rules for the generic structures will be explained later (see p. 93).

ODA document classes are not only used for specifying *structures* or common *properties* of specific documents belonging to a given document class. They can also be used for specifying *content*, or even more generally, any kind of information common to the documents of the document class. In other words, the concept of an ODA document class is used for "factorizing" certain kinds of information from the specific structures into the generic structures. In particular, content portions are permitted in the generic structures; these are then called *generic content portions*.

As a consequence, there may be ODA documents whose actual content is not completely described by the specific logical structure or specific layout structure. To display such documents on a rendition medium, the generic logical structure and generic layout structure also have to be taken into account. Since the generic structures have such a close relationship with the specific structures, the term *document* shall therefore be defined as follows: An ODA document consists of a specific logical structure, a specific layout structure, a generic logical structure and a generic layout structure. (Of course, some of these structures may not be present in a given document, as explained in the previous section for the documents without a logical or layout structure. The precise rules concerning the presence of the structures in certain situations are given in Sect. 3.3.)

For a better understanding consider once more the example given in Fig. 3. The "logo area" with the company logo in the associated content portion is a constituent of the specific layout structure. If a generic layout structure were associated with the document it would be rather natural from ODA's point of view to factorize this content portion into the generic layout structure and to have only a reference to this content portion within the specific layout structure. Thus one could avoid, for instance, storing the logo with each document belonging to the corresponding document class "business letter", thus reducing the storage space needed for the business letters of a company. On the other hand, to display such a letter on a rendition medium the generic layout structure would always be needed to supply the complete content of the letter.

It should be noted that the statement in Sect. 2.1.1, that a certain piece of content is associated with each basic object in the specific structures, has to be refined a little: this content is not necessarily specified directly within the specific structures but it may be associated indirectly with a basic object by reference to the generic structures.

2.1.3   The Document Processing Model

The ODA Standard was mainly developed for the electronic interchange of documents and therefore describes primarily the data structures of ODA documents and the concepts on which these data structures are based. The creation and processing of the documents is of secondary importance from the Standard's point of view. At several places, however, the Standard deals also with the so-called *document processing model*. This is necessary since ODA's architectural model implicitly makes some assumptions about the processing of ODA documents, and several concepts of the architectural model can only be understood when the processing model is known. The model is described briefly in this section with more details following later, especially in Sect. 3.6.

Concerning the processing of a specific document (not of a document class) the ODA Standard distinguishes the following three steps:

- the *editing process*,
- the *layout process* and
- the *imaging process*.

The *editing process* is concerned with creation and modification of the specific logical structure of a document. If the document belongs to a document class it may be necessary to take its generic logical structure into account to achieve a conforming specific logical structure.

The *layout process* creates a specific layout structure of a document from its specific logical structure. Again, if the document belongs to a document class the layout process will use information from the generic structures.

The *imaging process* creates the representation of a document on the rendition medium, for instance, on paper or on a computer screen, using the specific layout structure and the generic layout structure, if present. Until now it has been pretended that the specific layout structure of an ODA document completely describes its appearance on a rendition medium. However, this is not totally true. The physical properties of the rendition medium, for example, the resolution of a printer or the capability of a computer screen for rendering colours, might play a role for the final appearance of a document. The ODA Standard therefore introduces the concept of the imaging process to separate such hardware-dependent influences on the appearance of a document from the conceptual model of a "laid out" or "formatted" document.

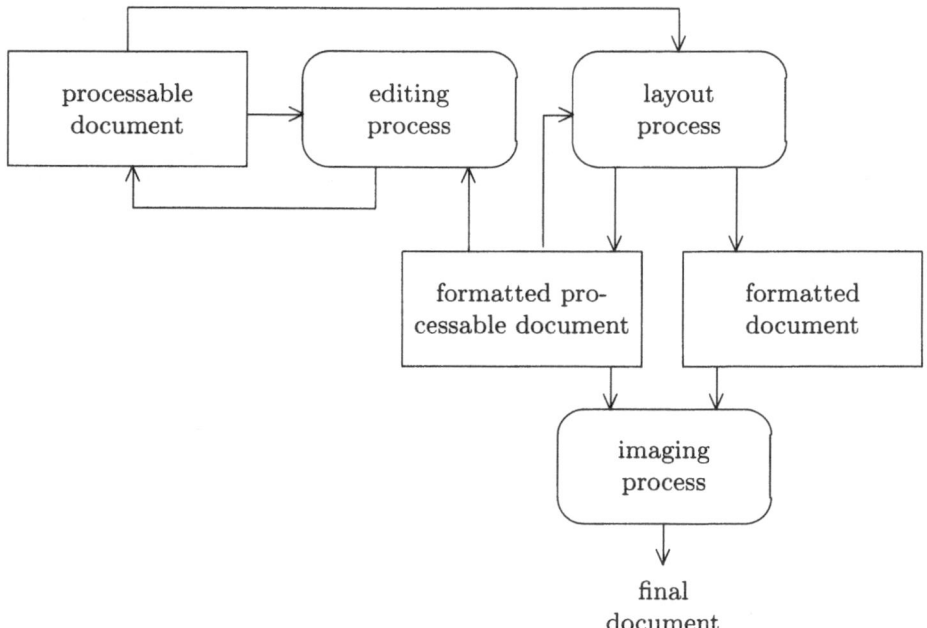

Fig. 5: Document processing model ODA

The processing model of ODA is shown graphically in Fig. 5. Three new terms are introduced in the figure, namely *processable document*, *formatted document* and *formatted processable document*.

A *processable document* is an ODA document with a specific logical structure and, possibly, a generic logical structure. The receiver of such a document can modify its content; however, he or she has to perform the layout process and imaging process before he or she can display the document on a rendition medium.

A *formatted document* is an ODA document with a specific layout structure and, possibly, a generic layout structure. The receiver of such a document can display it on a rendition medium after he has performed the (hardware dependent) imaging process; however, he cannot modify its content or appearance.

A *formatted processable document* is an ODA document with both a specific logical structure and a specific layout structure, and, possibly, generic structures. The receiver of such a document can display it on a rendition medium in the way specified by the sender using the layout structures. Furthermore, the receiver can modify the contents of the document using the logical structures. Afterwards he has to perform a layout process, of course, to generate a new layout structure reflecting the modifications, before the document can be displayed on a rendition medium.

In Fig. 5 the boxes with the rounded corners represent processes and the rectangles denote certain forms (processing stages) of a document. This means particularly that a *processable document*, a *formatted document* and a *formatted processable document* are not mere conceptual terms: the ODA Standard defines specific data structures for these three forms of a document; a document can be electronically interchanged in formatted, processable or formatted processable form. The arrows connecting the elements of the figure show which data structures are used or generated by each process.

The three processing steps (editing process, layout process and imaging process) are shown as separated processes in Fig. 5; however, for certain ODA implementations this processing model might not be recognizable from a user's point of view. For ODA implementations according to the WYSIWYG principle (*what you see is what you get*) where the user sees the document during editing always in the same form that the document will have when afterwards printed on paper, these three processes have to run simultaneously. The user will not be able to distinguish between the three processes. He only sees the *imaged form* of the document and, from his point of view, modifies the imaged form. The document processing system is responsible for making appropriate modifications to the specific logical structure and the specific layout structure simultaneously.

2.1.4   The Document Profile

Besides the constituents of an ODA document that were introduced above there is another constituent called document profile. It belongs neither to the specific structures nor to the generic structures of a document but is an object on its own. The document profile is specified in Part 4 of the Standard.

The document profile contains information about the the document to which it belongs; however, it includes no information which belongs to the actual content of the document. (However, some information in the document profile is required for processing the document.) For example, the document profile may contain information concerning the author of the document, the creation date or a list of the persons who received an electronic copy of the document. In particular, the document profile always specifies whether the document to which the document profile belongs is of formatted, processable or formatted processable form and which content architectures (character content, raster graphics content or geometric graphics content) appear in the document.

Each ODA document has a document profile. Therefore the term *document* shall be defined more precisely as follows: an ODA document consists of a document profile, a specific logical structure, a specific layout structure, a generic logical structure and a generic layout structure.

The ODA Standard explicitly provides for the electronic interchange of a document profile alone without the specific and generic structures of the document itself. The purpose for interchanging only the document profile is usually to test a receiving system's capability of processing the document to which the document profile belongs. This may sound a little bit strange since the aim of the ODA Standard is to facilitate the electronic interchange of documents conforming to the Standard, without requiring any additional agreements between sender and receiver. In principle, this is in fact correct: any systems implementing the (complete) ODA Standard should be able to receive and process any ODA document.

In reality, however, the situation may be more complicated. The Standard is quite comprehensive and a complete implementation therefore rather difficult and time consuming. For a continuing period it has to be expected that many systems will only implement a subset of the Standard; for instance, a system may not implement the raster graphics content architecture. For this reason, it was considered a useful feature of the Standard to permit the electronic interchange of a document profile alone, so that a receiving system, using the information contained in the document profile, may decide in advance whether or not it may be able to process the complete document to which the document profile belongs. (The issue of implementation restrictions will be discussed in greater detail in the following section.)

2.2   Document Application Profiles

The concept of document application profiles is based on the following rationale. The implementation of the complete ODA Standard, i.e., the development of a system which includes all concepts and features de-

fined in the Standard, is rather difficult and time consuming, due to
the great functionality of the Standard. In particular, such a system
must be able to process text, raster graphics and geometric graphics. In
many applications, however, a system implementing the complete Standard is not necessary. For business letters, for instance, it is usually
sufficient to process text; raster graphics or geometric graphics are not
needed.

It was therefore considered useful that developers and users of ODA
systems can specify subsets of the functionality of the Standard, according
to the requirements of certain classes of applications. Such a specification is done with document application profiles to which documents or
document classes can refer. (This is done with the attribute **document
application profile** in the document profile of a document.) When creating
a document which refers to a document application profile, only those
features can be used which are permitted by the document application
profile.

Furthermore, on almost all computers used in an office environment
there already exist systems for processing documents. It is not very realistic to assume that these systems will be replaced by ODA systems
without a transition phase. For a continuing period the users will wish
to use, or even have to use, both ODA systems and existing document
processing systems in parallel. In this case, document application profiles
can be used to define interfaces between ODA systems and existing systems, since many or even all features of a given system can be mapped to
ODA features; i.e., a document application profile can specify the features
which are supported by a given system, thereby describing the classes of
ODA documents it is able to process.

In a certain application, this might look as shown in Fig. 6.

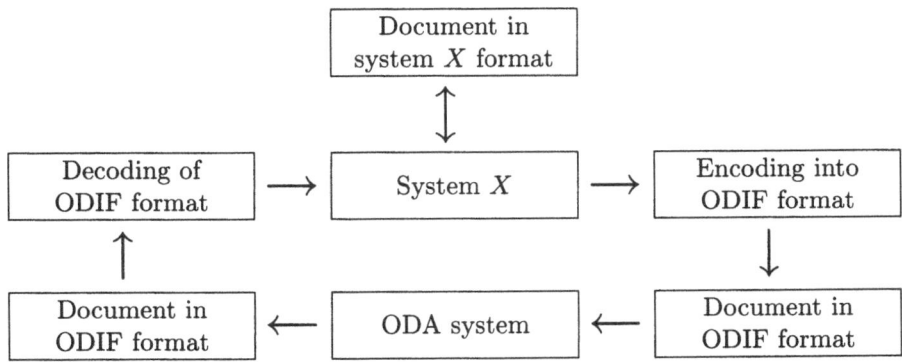

Fig. 6: Document interchange between ODA and non-ODA systems

When processing a document in system X the data format of this system is used. If the document is to be transmitted to an ODA system it is first encoded into the ODIF format so the ODA system can process it. At the receiving end, the ODIF format of a document created by an ODA system is decoded and converted to the data format of system X if the document is to be processed by this system.

It is even possible to go one step further and use the ODIF data format for document interchange between two non-ODA systems as shown in Fig. 7.

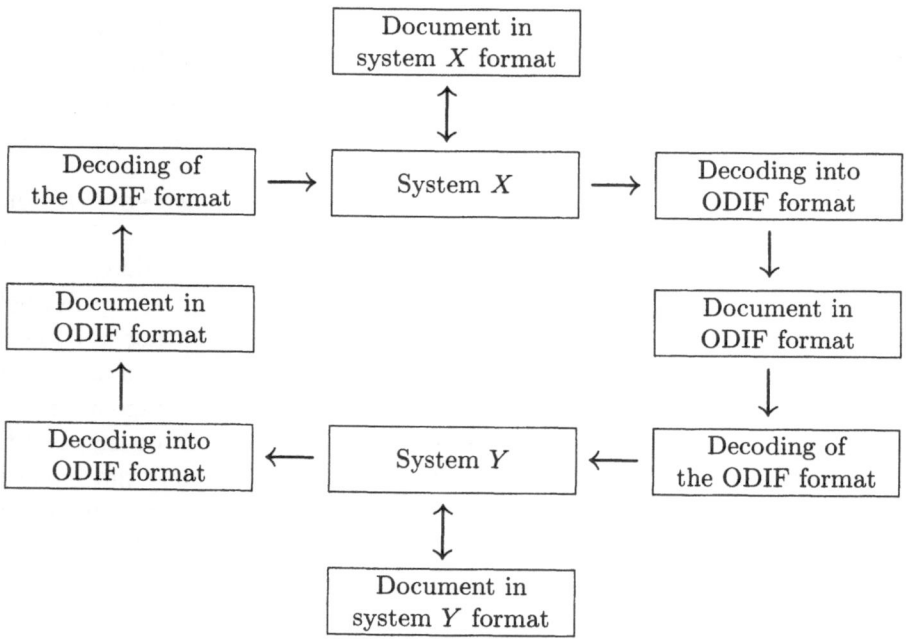

Fig. 7: Document interchange between two non-ODA systems using the ODIF format

In this application, the ODA Standard is only used to specify a common interface between two different non-ODA systems: ODIF is the interchange format and the concepts and features defined in the ODA Standard build a bridge between the concepts and features of system X and Y.

For a better understanding consider the following example: System X and system Y may both have the ability to center a piece of text within a line, for example, a chapter heading. The ODA Standard has also this feature and specifies how this is done in ODA documents (for instance, with the attribute block alignment). When encoding a document from the internal data format of system X into the ODIF format, ODA's way of specifying the centering of text must be used, which may be completely different from the method used by system X. When decoding the ODIF format into the internal format of system Y, ODA's specification method for centering must be recognized and translated into the way of system Y, which may be completely different both from the method of ODA and of system X.

It should be rather obvious that a given document processing system will hardly match the functionality of the ODA Standard (or of any other document processing system) completely. Usually, there will be only a certain subset of common features and this subset can be described by document application profiles, i.e., document application profiles can be used to describe certain classes of documents for which an interchange between an ODA system and a non-ODA system – or between two non-ODA systems using the ODIF encoding as an intermediate data format – is possible without loss of information.

In fact, software for converting between the ODIF format and the internal formats of existing document processing systems is currently being developed at many places. Furthermore, several national and international user groups of the ODA Standard have been established to develop document application profiles for their requirements.

From a rigorous point of view, the concept of document application profiles contradicts the main aim of the ODA Standard, namely to provide for the electronic interchange of documents without any additional agreements between the partners. When using document application profiles, negotiations between the partners are required in advance to define the content of the document application profiles, i.e., consensus must be reached on which subset of the functionality of the ODA Standard is permitted in the interchanged documents. The many activities concerning document application profiles which are currently going on indicate, however, a great interest in this concept. The vendors of document processing systems seem thereby to expect an integration, partial at least, of their products into the ODA world. The users of document processing systems apparently hope to solve at least some of their problems when interchanging documents between different systems.

2.2.1   The Content of Document Application Profiles

This section describes the specifications made in a document application profile. A detailed understanding of this section requires a previous reading of Chap. 3 ("Document Structures") since several terms will be used which are explained in that chapter.

When the ODA Standard was published in 1989, clause 9 of Part 1 contained a specification of the content of a document application profile. However, when user grcups actually started to develop document application profiles for their needs, it was discovered that this specification was not totally sufficient.

Therefore, the development of a so-called *Document Application Profile Proforma and Notation* was initiated. This Proforma and Notation is described in Sect. 2.2.3. For simplicity, the description of the document application profile in this section and the example of Sect. 2.2.2 are based on the original clause 9 of Part 1. It should be noted, however, that all current document application profile developments are based on the new Document Application Profile Proforma and Notation, which is not completely consistent with clause 9, though the general principles as explained below are identical.

A document application profile describes properties of ODA documents for which the document application profile applies. A document application profile can only restrict, but never extend the functionality of the ODA Standard, i.e., it always specifies a certain subset of the set of all valid ODA documents.

ODA documents and document classes are essentially defined by the specification of attributes and their values (see Sect. 3.1). Therefore, a document application profile specifies mainly the attributes and their values or value ranges which are permitted for a document conforming to the document application profile.

The following specifications have to be made in a document application profile, particularly:

– One or more so-called *document architecture levels* have to be specified. This includes a list of the permitted document architecture classes (formatted document architecture class, processable document architecture class and formatted processable document architecture class) and a definition the structures which can or must be present for each document architecture class, for instance, generic structures, specific structures or layout styles (see Sect. 3.3.1).

   For each of these structures, the permitted constituents must also be specified (see Sects. 3.3.1 and 3.3.2) as well as the restrictions which apply to these constituents. For example, it may be specified that frames and frame classes are not allowed within the layout structure and that each basic object can have only one associated content portion.

Furthermore, for each object or object class the set of the permitted attributes and their value ranges have to be defined. If more than one value is allowed for an attribute a so-called *basic value* and one or more so-called *non-basic values* have to be specified. A basic value is always permitted for the attribute whereas the usage of non-basic values must be announced in the document profile (see also p. 191). If an attribute is classified as defaultable its default value must be supplied.

For instance, it may be specified that the attribute bindings is not permitted for any object or object class, that the basic value for the attribute protection is unprotected, and that the basic value for the physical size of a page, given by the parameter nominal page size of the attribute medium type, is the size of an ISO A4 page, with any size between A5 and A3 being allowed as a non-basic value.

In addition, each attribute must be classified as mandatory, permitted (non-mandatory) or defaultable (see Sect. 3.2.4). The default values given for the attributes may be different from the default values specified in the Standard. Of course, only those values or value ranges can be specified which are contained in the value ranges defined by the Standard.

– One or more so-called *content architecture levels* have to be defined. This includes a list of the permitted content architectures (character content architecture, raster graphics content architecture and geometric graphics content architecture) with a specification of the permitted presentation attributes, control functions and coding attributes. Again, for each attribute its permissible values or value ranges, divided into basic and non-basic values, and for a defaultable attribute its default value must be supplied.

For example, a content architecture level of a document application profile may specify that only character content is allowed in a document, that the presentation attribute widow size must not be used and that the only permissible control functions are CR (carriage return), LF (line feed) and SP (space).

– A so-called *document profile level* has to be defined. This classifies each attribute which may appear in a document profile as either mandatory or permitted (non-mandatory) and defines permissible values or value ranges.

– The interchange format class used for the interchange of documents for which the document application profile applies must be specified.

Permitted interchange formats are ODIF, either class A or B, and SDIF (see Chap. 5).

2.2.2   Example of a Document Application Profile

Because of the great importance which document application profiles will probably have in the near future, an example of a document application profile is given in this section. A detailed understanding of the example requires a previous reading of Chap. 3 ("Document Structures") and Chap. 6 ("Character Content Architectures").

The document application profile introduced in this section describes a very simple subset of ODA documents: they contain only character content and are structured in a very simple manner. For the specification of the document application profile, a very simple notation will be used which does not conform to the Document Application Profile Proforma and Notation.

A document, belonging to the subset of ODA documents defined by the document application profile, shall be a formatted document, consisting only of a document profile, a specific layout structure and a presentation style. The layout structure contains only a document layout root, composite pages (subordinate to the document layout root) and blocks (subordinate to the composite pages). Only one content portion and the presentation style are associated with each block. The content portions shall contain only formatted character content. As a consequence, there is only one document architecture level and one content architecture level. The interchange format shall be ODIF, class A. The structure of the document application profile is therefore as shown in Fig. 8.

As a second step, the permissible attributes for each object of the specific layout structure have to be listed. The ODA Standard specifies that the attributes **object type** and **object identifier** are mandatory for the constituents document layout root, pages and blocks; the attribute **content identifier layout** must be specified for content portions. Furthermore, the attribute **subordinates** is required for the document layout root and for pages. In this example, also the attributes **presentation style** and **content portions** shall be specified for blocks because a presentation style and a content portion shall be associated with each block. A distinction between basic and non-basic values is not sensible for these attributes since there is always only one value possible.

In addition, it shall be allowed to specify the size of pages and blocks, i.e., the attribute **dimensions** shall be classified as permitted for these con-

Document Architecture Level:	
Permitted document architecture class:	formatted
Permitted structures:	specific layout structure
Permitted objects:	document layout root,
	composite pages (one or more),
	blocks (one or more),
	content portions (one or more),
	presentation style (one)
(additional specifications for the Document Architecture Level)	
Content Architecture Level:	
permitted content architecture class:	formatted character content
(additional specifications for the Content Architecture Level)	
Document Profile Level:	
(Specifications for the Document Profile Level)	
Interchange Format Class:	ODIF, format class A

Fig. 8: Structure of the document application profile

stituents. The basic value for a page shall be the size of an ISO A4 page and the basic value for a block shall be horizontally the width of the page and vertically the dimension which is required for the height of the text within the block. The non-basic values for pages shall be any page sizes between ISO A5 and A3. The non-basic values for blocks shall be specifications of the horizontal and vertical size of the block in scaled measurement units. Therefore, the specifications as shown in Fig. 9 must be made in the document architecture level of the document application profile.

Figure 9 list the attributes which may be specified for the objects of the specific layout structure. Each attribute is classified as either mandatory (m) or permitted (p) and its value or value range is defined. Except for the attribute object type, only a verbal description of the attribute values is given so the example is not complicated by syntactical details. (For instance, the value for the attribute content portions is described as "reference to a content portion". According to the ODA Standard, such a reference is in fact a sequence of non-negative integers (see p. 95); however, such mere syntactical details are not taken into account within the example of this section.)

For content portions the attributes content identifier layout (which is required for a unique identification of the content portions) and the attribute content information (which describes the actual content of the content portion) shall be mandatory; no other attributes shall be permitted. Within the text, i.e., within the value of the attribute content information,

Attributes for the document layout root:	
object identifier (m):	object identifier for the document layout root
object type (m):	document layout root
subordinates (m):	list of objects of type page

Attributes for pages:	
object identifier (m):	object identifier of a page
object type (m):	composite or basic page
subordinates (m):	list of objects of type block
dimensions (p):	basic value and default value: size of an A4 page
	non-basic values: all sizes between A5 and A3

Attributes for blocks:	
object identifier (m):	object identifier of a block
object type (m):	block
content portions (m):	reference to a content portion
presentation style (m):	reference to a presentation style
dimensions (p):	basic value and default value: horizontally the page size, vertically the height which is required according to the content of the block
	non-basic value: specification of the size of the block in scaled measurement units

Fig. 9: Required and permitted attributes for the objects of the specific layout structure

only characters from the minimum subrepertoire of ISO 6937-2 and the control functions SP (space), CR (carriage return) and LF (line feed) shall be allowed.

For the presentation style the attributes presentation style identifier and presentation attributes shall be mandatory; no other attributes shall be permitted. The value of the attribute presentation attributes consists of a list of the three presentation attributes character spacing, line spacing and first line offset which apply to character content. For these three attributes the basic, non-basic and default values have to be defined. According to the Standard these values have to be specified in basic measurement units or scaled measurement units, respectively. Therefore, the specifications for the content architecture level may be as shown in Fig. 10.

Finally it has to be specified which attributes may appear in the document profile of a document which conforms to the document application profile. In addition to the attributes document architecture class, content architecture classes, interchange format class, ODA version and document reference which are already classified as mandatory in the Standard, the attributes document application profile and specific layout structure shall

Attributes for content portions:	
content identifier layout (m):	content portion identifier
content information (m):	characters from the minimum subrepertoire of ISO 6937-2 and the specified control functions
Permitted control functions:	SP (space), CR (carriage return), LF (line feed)
Attributes for the presentation style:	
presentation style identifier (m):	identifier of a presentation style
presentation attributes (m):	list of the following three presentation attributes for formatted character content
character spacing (m):	basic value and default value: 120 BMU non-basic values: 100 BMU, 140 BMU
line spacing (m):	basic value and default value: 200 BMU non-basic values: 160 BMU, 180 BMU, 220 BMU
first line offset (m):	basic value and default value: 0 SMU non-basic values: all values between 1 SMU and 100 SMU

Fig. 10: Specifications for the content architecture level

be present, since a document application profile shall be associated with each document and each document shall have a specific layout structure. Furthermore, it shall be allowed to specify a default value for the page sizes in the document profile which can be done with the attribute **page dimensions**, i.e., the attribute **page dimensions** shall be classified as permitted and its value range has to be defined. In Fig. 11 the corresponding specifications for the document profile level are shown.

This document application profile example is very simple and the document application profiles which are currently the subject of international harmonization are much more complex. A document based on this document application profile consists only of text with characters from the character repertoire of ISO 6937-2. The pages of the document consist of simple paragraphs (blocks) and the only possibilities for influencing the layout of the documents are the selection of a few values for the character and line spacing and an indentation for the first line of a paragraph.

However, this very limited functionality should be common to almost all word processing systems in the office area, and converting such a simple structured document from the private format of a given system into the ODIF format and vice versa should be straightforward. This means that a document interchange between different word processing systems

Attributes of the document profile:

document architecture class (m):	formatted
content architecture classes (m):	⟦2 8 2 6 0⟧ (formatted character content)
interchange format class (m):	A
ODA version (m):	referenced version of the Standard
document reference (m):	identifier with which the document can be referenced
document application profile (m):	reference to the document application profile
specific layout structure (m):	present
page dimensions (p):	permitted values: all page sizes between A5 und A3

Fig. 11: Specifications of the document profile level

based on such a document application profile should always be possible, of course, at the expense of probably losing much of the functionality of a given system when converting a document into the ODIF format or vice versa.

2.2.3   The Document Application Profile Proforma and Notation

As mentioned above, the experiences gained when starting the actual specification of document application profiles led to the development of the Document Application Profile Proforma and Notation – abbreviated as DAP Proforma in the following. The Proforma was finished in the ISO committee responsible for the ODA Standard in late 1990 and will be published in 1991 as Annex F of ISO 8613, Part 1, and CCITT Recommendation T.411.

The purpose of this Annex F is to define a standardized format (Proforma) of document application profiles which is considered useful for registration authorities approving a profile, users, implementors and conformance testers.

The DAP Proforma requires that each document application profile contains the clauses shown in Fig. 12.

The main technical specifications of a particular document application profile are contained in its clauses 6 and 7. The specifications in clause 7 are to be given in a special notation which is defined in the DAP Proforma.

0	Introduction
1	Scope and field of application
2	References
3	Definitions and abbreviations
4	Relationships to other Document Application Profiles
5	Conformance
6	Characteristics supported by this Document Application Profile
	6.1 Overview
	6.2 Logical constituents
	6.3 Layout constituents
	6.4 Document layout control
	6.5 Content layout and imaging control
	6.6 Miscellaneous features
	6.7 Document managment features
7	Specification of constituent constraints
	7.1 Document profile constraints
	7.2 Logical constituent constraints
	7.3 Layout constituent constraints
	7.4 Layout style constraints
	7.5 Presentation style constraints
	7.6 Content portion constraints
	7.7 Additional usage constraints

Fig. 12: Outline of the Document Application Profile Proforma

In fact, the main part of the DAP Proforma consists of the definition of this notation.

The notation is based on the Backus-Naur-Form (BNF), comprising the syntactic and semantic definition of a set of terminal symbols and production rules. These terminal symbols and production rules are used in a particular document application profile for the specification of the constraints which apply to a document conforming to the profile.

The notation provides an unambiguous interpretation of the constraints. It is understandable by humans, in order to allow a derivation of the intended semantics, and since it uses a formal language, it is also machine readable and could therefore be used as a basis for the interchange of document application profile specifications between ODA systems.

A complete description of this notation cannot be given within this book but the example given in Fig. 13 may demonstrate some of its concepts.

The specifications in this example are to be interpreted as follows. A document conforming to this profile may contain objects and object

```
Passage:  {
GENERIC:
  REQ  Object-type                {'composite-logical-object'};
  REQ  Object-class-identifier    {ANY_VALUE};
  REQ  Generator-for-subordinates {$PassageGFS};
SPECIFIC:
  PERM Object-type                {'composite-logical-object'};
  REQ  Object-identifier          {ANY_VALUE};
  REQ  Object-class               {OBJECT_CLASS_ID_OF(Passage)};
  REQ  Subordinates               {SUB_ID_OF(NumberedSegment)+
                                   SUB_ID_OF(Paragraph)+
                                   SUB_ID_OF(BodyText)+
                                   SUB_ID_OF(BodyRaster)+
                                   SUB_ID_OF(BodyGeometric)+};
SPECIFIC_AND_GENERIC:
  PERM Layout-style               {STYLE_ID_OF(L-Style1)};
  PERM Bindings                   {$INITIALISEFOOTNOTE
                                   $INITIALISEANY};}
```

Fig. 13: Example of the Document Application Profile notation

classes of type "Passage". A corresponding object class must contain (REQ = required) the attributes object type, object class identifier and generator for subordinates. The value of the attribute object type is composite logical object, the value of the attribute object class identifier is any valid *object-class-id*, and the value of the attribute generator for subordinates is as defined by the macro "PassageGFS". This macro is defined somewhere else in the profile. A corresponding object must contain the attributes object identifier, object class and subordinates. The value of the attribute object identifier is any valid *object-id* and the value of the attribute object class must be an *object-class-id* of an object class of type "Passage". The value of the attribute subordinates is an ordered sequence of numbers, referring to one or more objects of the types "NumberedSegment", "Paragraph", "BodyText", "BodyRaster", and "BodyGeometric". These types of objects are defined somewhere else in the profile. The attribute object type may be specified (PERM = permitted) in which case its value is composite logical object. For an object class or an object of type "Passage" the attribute layout style and bindings may be specified. The value of the attribute layout style must identify a layout style of type "L-Style1" and the value of the attribute bindings is as specified by the macros INITIALISEFOOTNOTE and INITIALISEANY which are defined somewhere else in the profile.

In addition to the formal notation used in clause 7 of a particular document application profile, clause 6 of a profile shall provide a description of the characteristics of the constraints in a natural language form. Furthermore, any constraints which cannot be expressed with the formal notation should be described here. For instance, the BNF-based notation is in general insufficient for the specification of cross relations between attributes; such cross relations are therefore explained in clause 6.

The present state concerning the development of document application profiles is as follows. The specification of document application profiles was initiated by several user groups in Europe, North America and Asia, especially Japan, and within the CCITT. This work was internationally harmonized and resulted in essentially three different document application profiles: FOD 11 or CCITT Recommendation T.502, FOD 26 or CCITT Recommendation T.505, and FOD 36 or CCITT Recommendation T.506. The publication of the CCITT Recommendations is expected in 1991. The FOD profiles were submitted to ISO for registration as so-called *International Standardized Profiles* (ISPs); the ISP publication is expected in late 1991 or 1992.

These series of document application profiles have a hierarchical relationship, i.e., the Level 1 profiles (FOD 11 and T.502) are a subset of the Level 2 profiles (FOD 26 or T.505) which are in turn a subset of the Level 3 profiles (FOD 36 or T.506). Each level increases the functionality – or subset of ODA – supported by documents conforming to the profiles.

The Level 1 profiles support only character content with a rather limited degree of structure. The Level 2 profiles include raster graphics and geometric graphics with enhanced structuring capabilities and the Level 3 profiles support advanced document processing features.

2.3   Conformance

The ODA Standard defines conformance to ISO 8613 in terms of a data stream that represents an ODA document. There is a distinction between a document with and one without a referenced document application profile, i.e., between whether or not the attribute **document application profile** is specified in the document profile of the document.

If a document application profile is referenced, the document must satisfy the additional constraints given by the document application profile to be in conformance with the Standard. In any case, however, the document must fulfill the following conditions:

- The data stream has the format of one of the interchange format classes specified in Part 5 of the Standard.
- The data stream contains only the constituents described in Parts 2 and 4 with the restrictions concerning the combination of constituents, (for instance, for a specific document architecture class), the permitted attributes and their values.
- Only the content architectures described in Parts 6, 7 and 8 are used and only the attributes and their values defined in these parts appear in the data stream.
- Only character sets specified in ISO Standards or CCITT Recommendations are used.
- For geometric graphics the content elements conform to ISO 8632, the Computer Graphics Metafile Standard.
- The document, described by the constituents and their attributes, satisfies the specifications of the architectural model of ODA.

Basically, this definition of conformance requires that the elements in a data stream representing an ODA document must follow strictly the specifications in the Standard; deviations or extensions are not permitted. At a closer look, however, the conformance definitions are not completely precise. For example, by a mere static analysis of the data stream representing an ODA document it may not be possible to decide if the values of two attributes are compatible because the values of certain attributes are determined during the layout process. The layout process, however, is not described in full detail and may be subject to different interpretations. In this case it can become difficult to decide if a document satisfies the requirements of the Standard, i.e., whether it is a conforming ODA document.

It should be noted that the Standard is only concerned with the conformance of ODA *documents*, not of ODA *implementations*. This is correct from the Standard's point of view – its scope is the interchange, not the processing of documents – but it might not be totally acceptable from a user's point of view. A user intending to purchase an ODA system is surely interested whether the system implements the functionality of the Standard in agreement with the specifications. The term "conforming implementation", however, is not defined in the ODA Standard.

To fill this gap, ISO/IEC started a project to develop a *Technical Report on ISO 8613 Implementation Testing Methodology* which will be published as ISO/IEC TR 10183. The purpose of this report is to define a testing methodology and provide a framework for specifying abstract test cases for ISO 8613 implementation testing, the overall objective being the provision of a suitable base for testing the interworking capabilities of ODA implementations.

Part 1 of this Report, entitled *Framework*, defines terms used in the context of implementation testing, presents a conceptual model of ODA implementations, gives a phased approach to testing implementations, determines the functional components in generation and reception testing, and gives the requirements for abstract test cases.

Part 2, entitled *Abstract Test Suite*, specifies a framework for the development of abstract test cases and test documents, specifies a test case notation for abstract test cases, and gives examples of abstract test cases. In particular, this notation is based on the notation used for the specification of DAPs.

3 Part 2: Document Structures

This part of the ODA Standard has essentially the following technical content:

- The general concept for the data structures which are used for describing ODA documents is introduced.
- The constituents of the generic and specific structures and the cross-relations between them are specified.
- The interface between the document architecture and the content architectures is defined.
- A reference model for the document layout process and the document imaging process is described.
- A reference model for protecting parts of the document in respect to certain security aspects is specified.

This part does not contain information relating to specific content architectures (character, raster graphics or geometric graphics content). In other word, this part of the Standard is only concerned with the *structure* of documents, not with their actual *content*. The content of a document is enclosed in the content portions which are regarded essentially as "black boxes" whose internal structure is not considered in this part of the Standard.

3.1 Representation of the Document Structures

An introduction to ODA's architectural model was given in Sect. 2.1. For a representation of this conceptual model, especially for the electronic storage of documents conforming to this model, appropriate data structures are needed. For the specification of these data structures the Standard introduces the term *descriptive representation*.

The descriptive representation is based on the concept of a *constituent*. The structures in ODA's architectural model (the tree structures of the specific structures and the more general graph structures, in the sense of mathematical graph theory, of the generic structures) are modeled by these constituents where each constituent represents a node of the tree structures or graph structures. The descriptive representation of a document is thus a set (in the sense of mathematical set theory) of constituents.

The properties of a constituent and its relation to other constituents are described by so-called attributes, i.e., a constituent consists of a set of attributes where each attribute has a name and a value.

For a better understanding, the following example shall be used as the descriptive representation of a block. (Note: This is not the correct specification of a block but a gross simplification for tutorial purposes.)

"The descriptive representation of a block consists of the attributes **object type**, **object identifier**, **dimensions** and **content portions**. The value of the attribute **object type** is always **block**. The value of the attribute **object identifier** is a character string which represents a unique name for each block. The attribute **dimensions** has the two parameters **height** and **width** whose values are numbers specifying the vertical and horizontal dimensions of the block. The value of the attribute **content portions** is a non-empty set of character strings, each character string denoting the identifier of a content portion which is associated with the block."

This is basically the manner in which the descriptive representation of the constituents of ODA documents are defined in the Standard. (However, these specifications are often distributed over different places in the Standard and sometimes a little bit hard to find.) It should be rather obvious that this specification method only provides a more or less precise definition of the data *structures* of the constituents, but not a definition of the data *format*. This was done intentionally: the digital encoding of the data structures, i.e., the data format, can be chosen by the implementations according to their needs.

If, however, an ODA document is to be electronically interchanged, the data format must be defined up to the last bit, of course, so the receiver can interpret the data stream. This data format is called *Office Document Interchange Format* (ODIF) and defined in Part 5 of the Standard. Though an implementation could choose ODIF immediately for storing the descriptive representation of ODA documents, this is rather unlikely since ODIF is not very efficient for the processing of documents, for instance within an ODA editor.

If an ODA system were to choose Pascal as its implementation language, the descriptive representation of the block given in the example above could be realized with the following piece of code:

```
TYPE layout_object_type_value=(block,frame,page,
                               page_set,layout_root);
     dimensions_value=
        RECORD
           height: integer;
           width: integer;
        END;
     block_description=
        RECORD
           object_type: layout_object_type_value;
           object_identifier: string;
           dimensions: dimensions_value;
           content_portions: set of string;
        END;
VAR block_constituent: block_description;
    :
    :
    BEGIN ... block_constituent.object_type:=block; ...
```

(For simplicity, it was assumed that "**string**" is a predefined data type and "**set of string**" is a permitted construct. This is not the case with most Pascal compilers.)

Figure 14 shows the model for the descriptive representation of ODA documents. The constituents of a document are denoted by rectangles; sets which are built by collecting constituents, by rectangles with rounded corners.

In Fig. 14 five new terms appear which were not mentioned when introducing ODA's architectural model: *layout style* and *presentation style*, and the three superior terms *layout style set, presentation style set* and *style part*. In fact, these terms do not introduce any additional concepts to ODA's architectural model, which contains only the document profile, specific and generic structures, and logical and layout structures; these kinds of constituents were primarily introduced for mapping the architectural model onto data structures. The information represented by these constituents, however, cannot be assigned with either the logical structure alone or the layout structure alone, and therefore

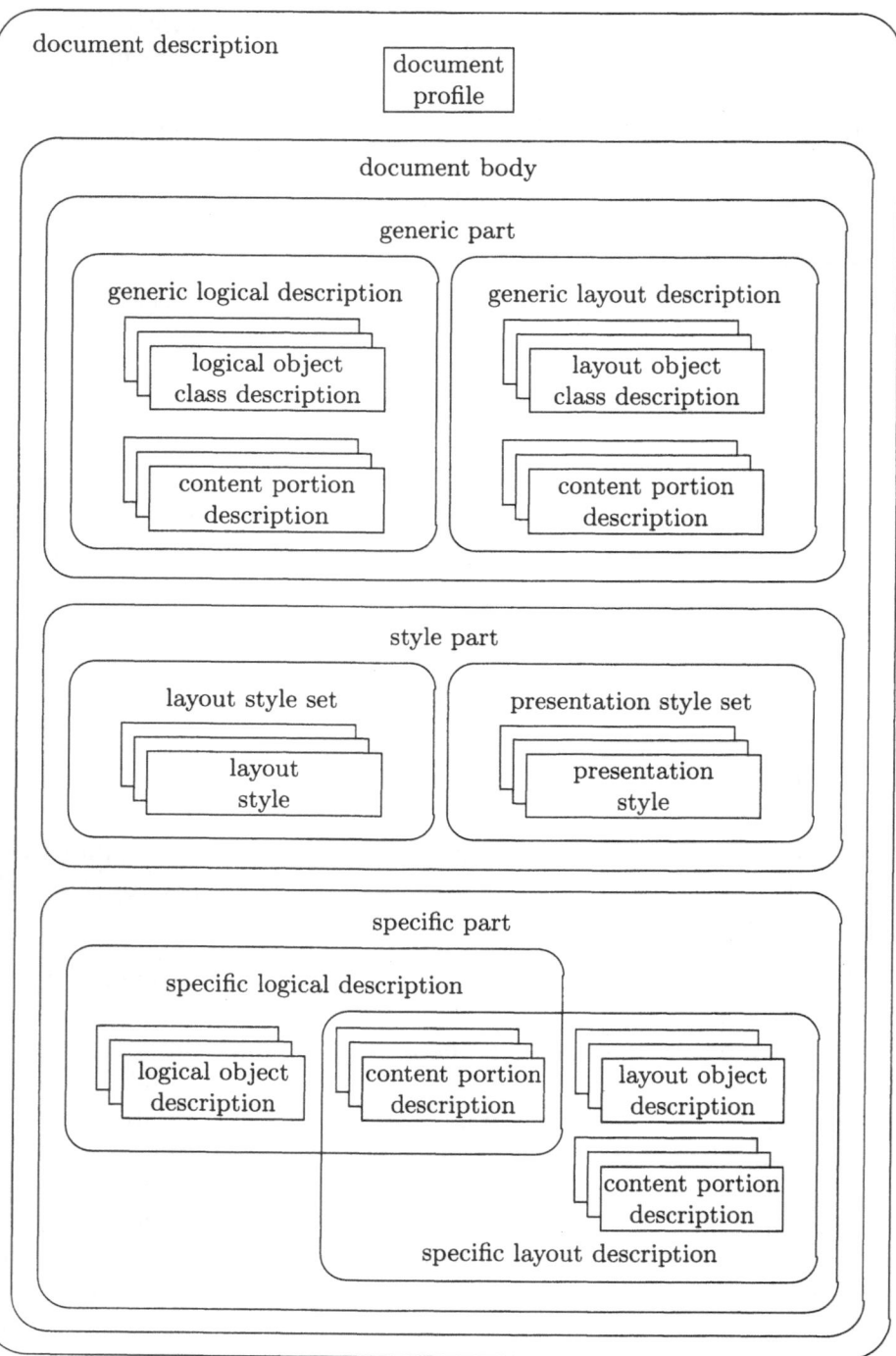

Fig. 14: Descriptive model of an ODA document

these new kinds of constituents were added to the descriptive representation.

Precisely speaking, a distinction must be made between the objects in ODA's architectural model (a concept) and the objects in ODA's descriptive representation (a data structure). This means that a generic logical structure is represented by a generic logical description, a logical object by a logical object description, or a frame by a frame description. This distinction, however, will usually not be made in the following, i.e., it will be said "A *frame* consists of the attributes ...", though the precise wording would be "A *frame description* consists of the attributes ...". Since the rest of the book is primarily concerned with the descriptive representation of ODA documents there is hardly any risk that this slightly imprecise wording will lead to ambiguities. (Even within the text of the Standard there is not always a clear distinction between an object and its descriptive representation.)

This original descriptive model of an ODA document as shown in Fig. 14 was extended by the Addendum on Security which added the so-called *protected document part* as shown in Fig. 15.

A protected document part consists of one or more so-called *sealed document profile descriptions, enciphered document profile descriptions, pre-enciphered document body part descriptions* and *post-enciphered document body part descriptions.*

At a closer look, the protected document part does not really add a new concept to the structure of ODA documents. It is rather a mechanism to encode certain parts of the document profile or the document body by special methods to deal with security aspects in the transmission of ODA documents as described in Sect. 3.5.

Simplifying slightly, the protected document part is created immediately before the originator of a document creates its ODIF data stream for interchange by moving certain constituents from other parts of the document into the protected part, applying an encryption technique to them. The intended receiver of a document, on the other hand, will move the constituents of the protected document part back to their original position when decoding the ODIF data stream, reversing the encryption applied to them. Therefore, for simplicity it may be assumed that the protected document part is only present in a document during its transmission but not during its processing on the originator's or receiver's side. (It will be explicitly noted in the following, whenever the presence of a protected document part is of importance for its processing.)

In the following sections, the constituents of a document will be introduced first and thereafter the attributes which are used for describing

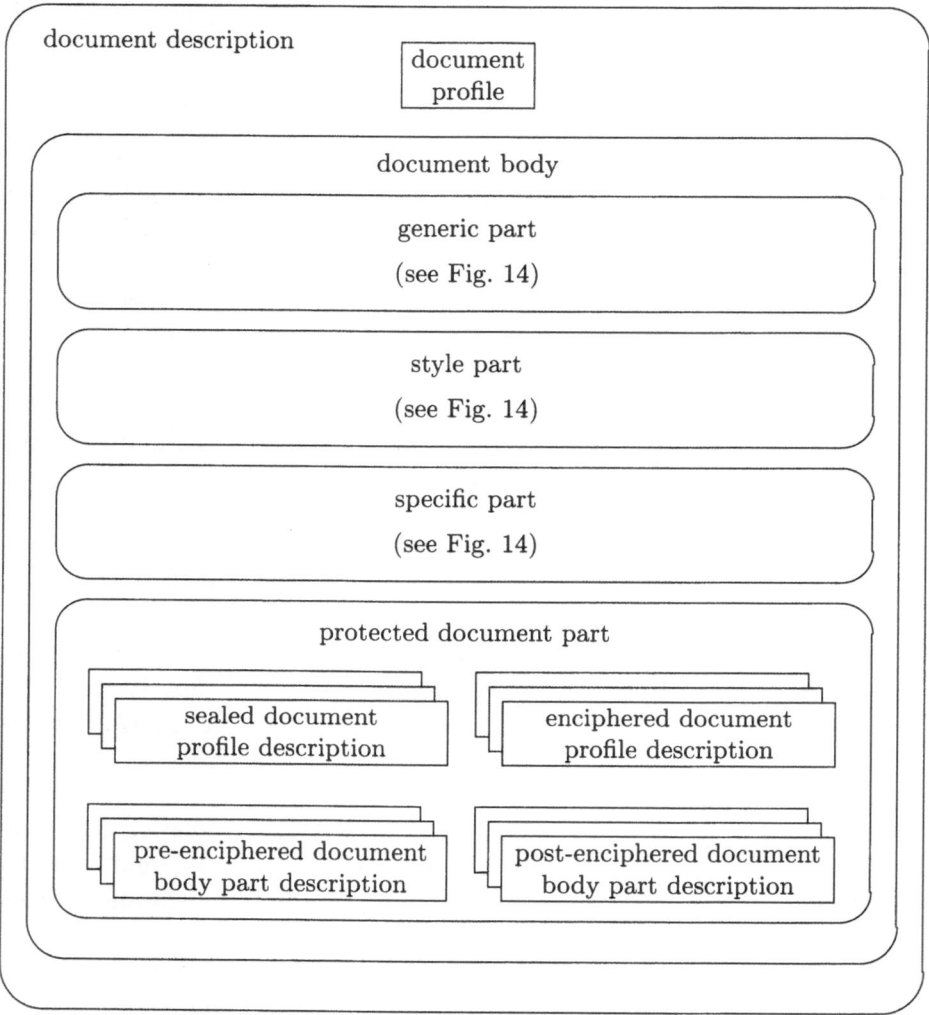

Fig. 15: Extended descriptive model of an ODA document including the protected document part

the constituents. (It should be kept in mind that a constituent is nothing else but a set of attributes describing the properties of the constituent and its relation to other constituents.)

Unfortunately, it cannot be avoided that the reader will have to look up the description of attributes at later places in the book when the constituents are described. The approach of describing an attribute in detail as soon as it appears with an attribute for the first time would have led to a rather confusing structure within this chapter.

The constituents shall be introduced according to the structures to which they belong. At first the constituents of the specific logical structure are described, then the constituents of the specific layout structure, afterwards the constituents of the generic logical and generic layout structure, then the content portions, layout styles and presentation styles and finally the constituents of the protected document part. The document profile, which is also a constituent of an ODA document, will be introduced in Chap. 4.

Before we start with the description of the constituents, we introduce a certain notation which will be used afterwards and might provide a clearer way for expressing the specifications of the Standard than the mere English language specifications of the Standard itself. The rules for this notation are as follows:

$a := \ldots$	denotes that the term a is defined by the specification "$\ldots$".
$\{\ldots\}$	denotes a set. (Example: $a := \{b, c\}$ specifies that a is a set with the two elements b and c.)
$[\ldots]$	denotes the optional presence of an object. (Example: "$[a]$" means that a may be present, but it does not need to be present.)
$\langle \ldots \rangle$	means that the object "$\ldots$" may be present, but if it is not it will be generated by a "defaulting mechanism". (This will be explained in more detail below.)
$a = \ldots$	means that a has the value "$\ldots$". This notation will be used for attributes and their values. (Example: "object type = block" specifies that the attribute with the name "object type" has the value "block".)
$\langle a = \ldots \rangle$	denotes that the attribute a is optional, however, if it is present it must have the value "$\ldots$".

To understand the notation "$\langle \ldots \rangle$" it should be known that ODA distinguishes between three classes of attributes of a constituent. There are attributes (usually depending on the kind of constituent) which must always be specified, called *mandatory attributes*, attributes which may optionally be specified for a constituent, called *non-mandatory attributes*, and attributes for which a default value will be determined according to a defaulting mechanism, if they are not explicitly specified for a constituent, called *defaultable attributes*.

In principle, defaultable attributes can be regarded as mandatory attributes: though their values may not be specified directly within the attribute set of the constituent, their values can be found "somewhere else". (The mechanism for determining the value of defaultable attributes is explained in Sect. 3.2.16.)

Consider, for instance, a layout object of type block which represents a rectangle on the presentation medium and therefore has a certain height and width, specified by the attribute dimensions. If a block belongs to a certain block class in a given ODA document and the attribute dimensions is specified for the block class, the attribute dimensions may not necessarily be specified for the block explicitly, but the default value for the size of the block will be derived from the corresponding specifications in the block class. (The actual rules for determining the default value are more complicated, in fact.)

3.1.1   The Constituents of the Specific Logical Structure

The constituents of the specific logical structure are the document logical root, composite logical objects and basic logical objects.

Document logical root

A document logical root is represented by the following attribute set:

document logical root :=
  {object identifier, ⟨object type = document logical root⟩,
  subordinates,
  ⟨application comments⟩, ⟨bindings⟩, ⟨protection⟩, ⟨sealed⟩,
  ⟨user-readable comments⟩, ⟨user-visible name⟩,
  [default value lists], [enciphered], [layout style], [object class]}

The attribute object identifier must always be specified, also for the other constituents of the specific structure. This is rather obvious since the value of this attribute is essentially the "name" of the constituent, and each constituent must have a unique name by which a reference can be made from other constituents and by which objects of the same kind can be distinguished. (In contrast to the other types of objects in the logical structure, however, there can be only one document logical root.)

The attribute object type need not be specified if the attribute object class is present, but if it it specified its value must be document logical root. Correspondingly, the value of the attribute object class, if specified, must refer to a document logical root class (see the description of this attribute on p. 96). Since this relation between these two attributes is also valid for other constituents, it shall be recorded for later reference purposes:

Note 1:   The attribute **object type** may be missing if the attribute object class is present. In this case, the type of the constituent referred to by the attribute **object class** determines the value of the attribute **object type**.

The attribute **subordinates** which denotes the objects immediately subordinate to the document logical root must also be specified, and the constituents which are referred to by this attribute must be composite logical objects or basic logical objects.

Though the attribute **object class** is non-mandatory in principle (see, however, Note 1), it must be specified for documents whose generic logical structure is a so-called *complete generator set*. (The term *complete generator set* is explained in Sect. 3.2.15.) Since this is also valid for other constituents it shall be recorded for later reference purposes:

Note 2:   The attribute **object class** must be specified if the generic logical structure is a complete generator set (see Sect. 3.2.15).

The other attributes of the document logical root will be explained later; see the descriptions of the attributes on p. 109 (application comments), 102 (bindings), 138 (default value lists), 136 (enciphered), 97 (layout style), 96 (object class), 88 (object identifier), 88 (object type), 137 (protection), 136 (sealed), 91 (subordinates) and 108 (user-readable comments and user-visible name).

It should be noted that the attributes **enciphered** and **sealed** were introduced by the Addendum on Security; they were not contained in the initially published version of the Standard.

Composite logical object

A composite logical object is represented by the following attribute set:

composite logical object :=
    {object identifier, ⟨object type = composite logical object⟩,
    subordinates,
    ⟨application comments⟩, ⟨bindings⟩, ⟨protection⟩, ⟨sealed⟩,
    ⟨user-readable comments⟩, ⟨user-visible name⟩,
    [default value lists], [enciphered], [layout style], [object class]}

The attribute set is identical to the one for the document logical root, except that the value of the attribute **object type**, if present, must be com-

posite logical object. Note 1 holds for the relation between the attributes
object type and **object class**. The constituent which may be referred to by
the attribute **object class** must be a composite logical object class. Note 2
is also valid.

The attribute **subordinates** which denotes the objects immediately subordinate to the composite logical object must also be specified, and the
constituents which are referred to by this attribute must be composite logical objects or basic logical objects. Therefore, there can be any number
$(0, 1, 2, \ldots n)$ of hierarchical levels between the document logical root and
a certain basic logical object. The structure of the value of the attribute
subordinates guarantees that the objects of the specific logical structure
build a tree structure (see p. 92), i.e., a composite logical object can never
be subordinate to itself, either directly or indirectly.

The other attributes of composite logical objects will be explained
later; see the descriptions of the attributes on p. 109 (application comments), 102 (bindings), 138 (default value lists), 136 (enciphered), 97 (layout
style), 96 (object class), 88 (object identifier), 88 (object type), 137 (protection), 136 (sealed), 91 (subordinates) and 108 (user-readable comments
and user-visible name).

It should be noted that the attributes **enciphered** and **sealed** were introduced by the Addendum on Security; they were not contained in the
initially published version of the Standard.

Basic logical object

A basic logical object is represented by the following attribute set:

> *basic logical object* :=
>   {object identifier, ⟨object type = basic logical object⟩,
>   ⟨application comments⟩, ⟨bindings⟩, ⟨content architecture class⟩,
>   ⟨protection⟩, ⟨sealed⟩, ⟨user-readable comments⟩, ⟨user-visible name⟩,
>   [alternative], [content generator], [content portions], [enciphered],
>   [layout style], [object class], [primary], [presentation style]}

In contrast to composite logical objects, the attribute **subordinates** is
not permitted, of course. The value of the attribute **object type**, if present,
is **basic logical object**. Note 1 holds for the relation between the attributes
object type and **object class**. The constituent which may be referred to by
the attribute **object class** must be a basic logical object class. Note 2 is
also valid.

The attribute **default value lists** is not permitted for basic logical objects since it is only sensible for constituents with subordinate objects (see the explanation of this attribute on p. 138).

The attribute **primary** is only applicable if the basic logical object is a so-called *alternative description*. In this case, the attribute is mandatory.

Compared to composite logical objects, the attributes **alternative**, **content architecture class**, **content generator**, **content portions**, **primary** and **presentation style** have been added. Each basic logical object is associated with a certain piece of the actual content of the document and each piece of content has a specific content architecture. Therefore, the attribute **content architecture class** must have a value which is determined by the defaulting mechanism if the attribute is not specified directly with the basic logical object.

The attribute **content portions** may be missing, but in this case the content associated with the basic logical object will be derived from the attribute **content generator** or from the object class to which the basic logical object refers. The precise algorithm for determining the content associated with basic logical objects is described in Sect. 3.6.2, p. 164ff.

The other attributes of the basic logical objects will be explained later; see the descriptions of the attributes on p. 89 (alternative), 109 (application comments), 102 (bindings), 99 (content architecture class), 103 (content generator), 95 (content portions), 136 (enciphered), 97 (layout style), 96 (object class), 88 (object identifier), 88 (object type), 97 (presentation style), 90 (primary), 137 (protection), 136 (sealed) and 108 (user-readable comments and user-visible name).

It should be noted that the attributes **alternative** and **primary** were introduced by the Addendum on Alternate Representations and the attributes **enciphered** and **sealed** by the Addendum on Security; they were not present in the initially published version of the Standard.

3.1.2   The Constituents of the Specific Layout Structure

The constituents of the specific layout structure are the document layout root, page sets, pages, frames and blocks.

Document layout root

A document layout root is represented by the following attribute set:

document layout root :=
  {object identifier, ⟨object type = document layout root⟩,
    subordinates,
    ⟨application comments⟩, ⟨balance⟩, ⟨bindings⟩, ⟨sealed⟩,
    ⟨user-readable comments⟩, ⟨user-visible name⟩,
    [default value lists], [enciphered], [object class]}

The attributes object identifier and subordinates must usually be specified and therefore these attributes are classified as mandatory in the definition above. However, there is a certain type of ODA document which shall be called "*simple-structured CCITT document*" (see Sect. 3.3.3) for which these attributes may be missing. Since this type of document violates slightly the general principles of ODA's architectural model, it may be assumed for simplicity that the attributes object identifier and subordinates must always be specified for a document layout root. For later reference purposes, however, it shall be recorded:

Note 3:   The attribute object identifier is non-mandatory for simple-structured CCITT documents (see Sect. 3.3.3).

Note 4:   The attribute subordinates is non-mandatory for simple-structured CCITT documents (see Sect. 3.3.3).

Concerning the relation between the attributes object type and object class, Note 1 holds again. The constituent which may be referred to by the attribute object class must be a document layout root class. Similar to Note 2 it shall be recorded:

Note 5:   The attribute object class must be specified if the generic layout structure is a complete generator set (see Sect. 3.2.15).

For a document layout root, the value of the attributes refers to constituents which are page sets or pages. A mixture of page sets and pages is permitted.

The other attributes of the document layout root will be explained later; see the descriptions of the attributes on p. 109 (application comments), 125 (balance), 102 (bindings), 138 (default value lists), 136 (enciphered), 96 (object class), 88 (object identifier), 88 (object type), 136 (sealed), 91 (subordinates) and 108 (user-readable comments and user-visible name).

It should be noted that the attributes enciphered and sealed were introduced by the Addendum on Security; they were not contained in the initially published version of the Standard.

Page set

A page set is represented by the following attribute set:

page set :=
 {object identifier, ⟨object type = page set⟩,
  subordinates,
  ⟨application comments⟩, ⟨balance⟩, ⟨bindings⟩, ⟨sealed⟩,
  ⟨user-readable comments⟩, ⟨user-visible name⟩,
  [default value lists], [enciphered], [object class]}

Note 1 holds for the relation between the attributes object type and object class. The constituent which may be referred to by the attribute object class must be a page set class. Note 5 is also valid.

The value of the attribute subordinates refers to constituents which are page sets or pages; a mixture of page sets and pages is permitted. Therefore, there can be any number $(0, 1, 2, \ldots n)$ of hierarchical levels of page sets between the document layout root and specific page. However, it is not possible that a given page set has itself as a subordinate object, either directly or indirectly.

The other attributes of page sets will be explained later; see the descriptions of the attributes on p. 109 (application comments), 125 (balance), 102 (bindings), 138 (default value lists), 136 (enciphered), 96 (object class), 88 (object identifier), 88 (object type), 136 (sealed), 91 (subordinates) and 108 (user-readable comments and user-visible name).

It should be noted that the attributes enciphered and sealed were introduced by the Addendum on Security; they were not contained in the initially published version of the Standard.

Pages

In principle, ODA distinguishes between two types of pages: *composite pages* and *basic pages*. In particular, the attribute sets which can be specified for these two kind of pages are different. Therefore, composite pages and basic pages shall be introduced as different types of constituents.

It should be noted, however, that the attribute object type has the same value for both types of pages, namely composite or basic page, i.e., composite pages and basic pages cannot be distinguished by the value of this attribute. Whether a page is composite or basic can usually be decided by the presence of the attribute subordinates: basic pages have no subordinate objects. (There is an exception to this rule as will be explained later.)

(This way of determining the type of a constituent is a special case in ODA's descriptive representation since for all other constituents their type is defined by the attribute object type. It would be more logical if the attribute object type had different values for composite and basic pages.)

Furthermore, it should be noted that the specific layout structure of an ODA document can contain only composite pages or basic pages; a mixture of both is not permitted.

Composite page

A *composite page* is represented by the following attribute set:

composite page :=
  {object identifier, ⟨object type = composite or basic page⟩,
  subordinates,
  ⟨application comments⟩, ⟨balance⟩, ⟨bindings⟩, ⟨colour⟩,
  ⟨dimensions⟩, ⟨medium type⟩, ⟨page position⟩, ⟨sealed⟩,
  ⟨transparency⟩, ⟨user-readable comments⟩, ⟨user-visible name⟩,
  [default value lists], [enciphered], [imaging order], [object class]}

Note 1 holds for the relation between the attributes object type and object class. The constituent which may be referred to by the attribute object class must be a composite page class. Note 5 is also valid.

The value of the attribute subordinates refers to constituents which are either frames or blocks; a mixture of frames and blocks is not allowed.

The attribute balance which is classified as non-mandatory is only sensible for composite pages whose immediate subordinates are frames. Therefore, if the value of the attribute subordinates refers to constituents of type block, the attribute balance must not be specified and no default value will be determined. If, however, the value of the attribute subordinates refers to constituents of type frame, the attribute balance may be specified; otherwise a default value will be determined.

(Concerning the simple-structured CCITT documents, Notes 3 and 4 hold again. This immediately raises the question of how composite pages and basic pages can be distinguished for these documents since the presence of the attribute subordinates is usually their discrimination criterion. Answer: If such a document contains blocks the pages are composite pages, otherwise basic pages.)

Within the attribute set of composite pages there appear also several attributes related to the physical appearance of this object on a rendition medium, namely colour, dimensions, medium type, page position, transparency and imaging order. However, these (and the other) attributes shall not be considered in detail here; see the descriptions of the attributes on p. 109 (application comments), 125 (balance), 102 (bindings), 134 (colour), 138 (default value lists), 119 (dimensions), 136 (enciphered), 132 (imaging order), 135 (medium type), 96 (object class), 88 (object identifier), 88 (object type), 134 (page position), 136 (sealed), 91 (subordinates), 133 (transparency) and 108 (user-readable comments and user-visible name).

It should be noted that the attributes enciphered and sealed were introduced by the Addendum on Security; they were not contained in the initially published version of the Standard.

Basic page

A *basic page* is represented by the following attribute set:

> *basic page* :=
>   {object identifier, ⟨object type = composite or basic page⟩,
>   ⟨application comments⟩, ⟨bindings⟩, ⟨colour⟩,
>   ⟨content architecture class⟩, ⟨dimensions⟩, ⟨medium type⟩,
>   ⟨page position⟩, ⟨presentation attributes⟩, ⟨sealed⟩, ⟨transparency⟩,
>   ⟨user-readable comments⟩, ⟨user-visible name⟩,
>   [alternative], [content portions], [enciphered], [object class], [primary],
>   [presentation style]}

Note 1 holds for the relation between the attributes object type and object class. The constituent which may be referred to by the attribute object class must be a basic page class. Concerning the simple-structured CCITT documents, the same comments as made for composite pages are valid; in particular, Notes 3 and 4 hold again.

Compared to composite pages, the attributes subordinates and balance are missing; the attributes alternative, content architecture class, content portions, primary, presentation attributes and presentation style have been added.

The attribute primary is only applicable if the basic page is a so-called *alternative description*. In this case, the attribute is mandatory.

Each basic page is associated with a certain piece of the actual content of the document and this piece of content has a certain content architecture. Therefore, the attribute content architecture class must be specified or a default value will be derived.

The attribute content portions may be missing but then the content associated with the basic page will be derived from the basic page class to which it refers (see Sect. 3.3.3). Concerning the simple-structured CCITT documents, the attribute may be missing even if there are content portions associated with the basic page (see Sect. 3.3.3).

The other attributes of basic pages will be explained later; see the descriptions of the attributes on p. 89 (alternative), 109 (application comments), 102 (bindings), 134 (colour), 99 (content architecture class), 95 (content portions), 119 (dimensions), 136 (enciphered), 135 (medium type), 96 (object class), 88 (object identifier), 88 (object type), 134 (page position), 101 (presentation attributes), 97 (presentation style), 90 (primary), 136 (sealed), 133 (transparency) and 108 (user-readable comments and user-visible name).

It should be noted that the attributes alternative and primary were introduced by the Addendum on Alternate Representations and the attributes enciphered and sealed by the Addendum on Security; they were not present in the initially published version of the Standard.

Frame

A frame is represented by the following attribute set:

frame :=
  {object identifier, ⟨object type = frame⟩,
  subordinates,
  ⟨application comments⟩, ⟨balance⟩, ⟨bindings⟩, ⟨border⟩,
  ⟨colour⟩, ⟨dimensions⟩, ⟨layout path⟩, ⟨permitted categories⟩,
  ⟨position⟩, ⟨sealed⟩, ⟨transparency⟩, ⟨user-readable comments⟩,
  ⟨user-visible name⟩,
  [default value lists], [enciphered], [imaging order], [object class]}

Note 1 holds for the relation between the attributes object type and object class. The constituent which may be referred to by the attribute object class must be a frame class. Note 5 is also valid.

The value of the attribute subordinates refers to constituents which are either frames or blocks; a mixture of frames and blocks is not allowed. If blocks appear as subordinate objects such a frame is called a *lowest level frame*. Therefore, there may be any number $(0, 1, 2, \ldots n)$ of hierarchical levels of frames between a page and a certain block. The structure of the value of the attribute subordinates guarantees that the objects of the specific layout structure build a tree structure (see p. 92), i.e., a frame can never be subordinate to itself, neither directly nor indirectly.

The attribute balance which is classified as defaultable must not be specified for *lowest level frames* (frames whose immediate subordinates are blocks) and no default value will be determined for such frames.

The attribute permitted categories is only permitted for lowest level frames.

The other attributes of frames will be explained later; see the descriptions of the attributes on p. 109 (application comments), 125 (balance), 102 (bindings), 121 (border), 134 (colour), 138 (default value lists), 119 (dimensions), 136 (enciphered), 132 (imaging order), 110 (layout path), 96 (object class), 88 (object identifier), 88 (object type), 128 (permitted categories), 117 (position), 136 (sealed), 91 (subordinates), 133 (transparency) and 108 (user-readable comments and user-visible name).

It should be noted that the attributes enciphered and sealed were introduced by the Addendum on Security; they were not contained in the initially published version of the Standard.

Block

A block is represented by the following attribute set:

block :=
   {object identifier, ⟨object type = block⟩,
    ⟨application comments⟩, ⟨bindings⟩, ⟨border⟩, ⟨colour⟩,
    ⟨content architecture class⟩, ⟨dimensions⟩, ⟨position⟩,
    ⟨presentation attributes⟩, ⟨sealed⟩, ⟨transparency⟩,
    ⟨user-readable comments⟩, ⟨user-visible name⟩,
    [alternative], [content portions], [enciphered], [object class], [primary],
    [presentation style]}

Note 1 holds for the relation between the attributes object type and object class. The constituent which may be referred to by the attribute object class must be a block class. Concerning the simple-structured CCITT documents, Note 3 holds again.

Each block is associated with a certain piece of the actual content of the document and this piece of content has a certain content architecture. Therefore, the attribute content architecture class must be specified or a default value will be derived.

The attribute primary is only applicable if the block is a so-called *alternative description*. In this case, the attribute is mandatory.

The attribute content portions may be missing but then the content associated with the block will be derived from the block class to which it refers (see Sect. 3.3.3). Concerning the simple-structured CCITT documents, the attribute may be missing even if there are content portions associated with the block (see Sect. 3.3.3).

The other attributes of basic pages will be explained later; see the descriptions of the attributes on p. 89 (alternative), 109 (application comments), 102 (bindings), 121 (border), 134 (colour), 99 (content architecture class), 95 (content portions), 119 (dimensions), 136 (enciphered), 96 (object class), 88 (object identifier), 88 (object type), 117 (position), 101 (presentation attributes), 97 (presentation style), 90 (primary), 136 (sealed), 133 (transparency) and 108 (user-readable comments and user-visible name).

It should be noted that the attributes alternative and primary were introduced by the Addendum on Alternate Representations and the attributes enciphered and sealed by the Addendum on Security; they were not present in the initially published version of the Standard.

3.1.3   The Constituents of the Generic Logical Structure

The constituents of the generic structures are built very similarly to their corresponding constituents in the specific structures, i.e., all attributes which appear with a certain object in the specific structures appear also with the corresponding object class in the generic structures. To this principle, there are only the following exceptions:

- Instead of the attribute object identifier in the specific structures there is the attribute object class identifier in the generic structures.
- Instead of the attribute subordinates in the specific structures the attribute generator for subordinates is used in the generic structures.
- The attribute object class by which an object in the specific structures refers to an object class in the generic structures is not present with object classes.
- The attribute imaging order (see p. 132) appears only within the specific structure, namely with composite pages and frames.

- the attributes resource (see p. 98) and logical source (see p. 106) can only be used for constituents of the generic structures (resource for all constituents of the generic structures, logical source only for frame classes).
- The attribute content generator (see p. 103) is allowed for basic logical object classes, basic page classes and block classes in the generic structures but only for basic logical objects in the specific structures.

Within the generic structures, the attributes object class identifier and object type must always be specified, i.e., they are mandatory attributes for object classes.

Within the generic structures, there are only mandatory and non-mandatory attributes. Defaultable attributes are obviously not sensible within the generic structures, since "above" the generic structures there are no other structures from which a default value could be derived. (However, the ODA Standard defines for many attributes a kind of "predefined" default value which is used when a value for a defaultable attribute can be found neither in the specific structures nor in the generic structures.)

The constituents of the generic logical structure are the document logical root class, composite logical object classes and basic logical object classes.

Document logical root class

A document logical root class is represented by the following attribute set:

document logical root class :=
  {object class identifier, object type = document logical root,
  [application comments], [bindings], [default value lists],
  [generator for subordinates], [layout style], [protection], [resource],
  [sealed], [user-readable comments], [user-visible name]}

The attribute generator for subordinates usually specifies which objects can be immediately subordinate to a document logical root belonging to the document logical root class (see description of the attribute on p. 93). (An exception to this rule will be described later.) Since a document logical root always has subordinate objects the classification of this attribute as non-mandatory might be a little bit surprising. This is again due to the different kinds of generic structures which may appear in an ODA document: If the generic logical structure is a complete generator set (see Sect. 3.2.15) the attribute must be specified, i.e., in this case it is a mandatory attribute. If the generic logical structure is a so-called *factor set* (see Sect. 3.2.15) the attribute must not be specified.

For later reference purposes this shall be recorded as

Note 6:   The attribute **generator for subordinates** must be specified if the generic logical structure of a document is a complete generator set; it is not allowed if the generic logical structure is a factor set (see Sect. 3.2.15).

Besides complete generator sets and factor sets there is a third kind of generic logical structures, called *partial generator sets.* For partial generator sets the attribute **generator for subordinates** has a weaker "control" over the subordinate objects of a document logical root belonging to the document logical root class: For partial generator sets there may be objects immediately subordinate to a document logical root which are not explicitly permitted by the attribute **generator for subordinates** of the respective document logical root class (see the description of the attribute on p. 93 for more details).

For later reference purposes this shall be recorded as

Note 7:   If the generic logical structure of a document is a partial generator set (see Sect. 3.2.15) the objects in the specific logical structure belonging to the object class may have subordinate objects which are not explicitly specified by the attribute **generator for subordinates**.

The other attributes of a document logical root class will be explained later; see the descriptions of the attributes on p. 109 (**application comments**), 102 (**bindings**), 138 (**default value lists**), 93 (**generator for subordinates**), 97 (**layout style**), 88 (**object class identifier**), 88 (**object type**), 137 (**protection**), 98 (**resource**), 136 (**sealed**) and 108 (**user-readable comments and user-visible name**).

It should be noted that the attribute **sealed** was introduced by the Addendum on Security; it was not contained in the initially published version of the Standard.

Composite logical object class

A composite logical object class is represented by the following attribute set:

composite logical object class :=
  {object class identifier, object type = composite logical object,
  [application comments], [bindings], [default value lists],
  [generator for subordinates], [layout style], [protection], [resource],
  [sealed], [user-readable comments], [user-visible name]}

The attribute set is identical to the one for a document logical root class; however, the value of the attribute object type must be composite logical object. Concerning the attribute generator for subordinates, Notes 6 and 7 hold.

The description of the attributes can be found on p. 109 (application comments), 102 (bindings), 138 (default value lists), 93 (generator for subordinates), 97 (layout style), 88 (object class identifier), 88 (object type), 137 (protection), 98 (resource), 136 (sealed) and 108 (user-readable comments and user-visible name).

It should be noted that the attribute sealed was introduced by the Addendum on Security; it was not contained in the initially published version of the Standard.

Basic logical object class

A basic logical object class is represented by the following attribute set:

basic logical object class :=
  {object class identifier, object type = basic logical object,
  [application comments], [bindings], [content architecture class],
  [content generator], [content portions], [enciphered], [layout style],
  [presentation style], [protection], [resource], [sealed],
  [user-readable comments], [user-visible name]}

Compared to a composite logical object class the attributes generator for subordinates and default value lists are missing and the attributes content architecture class, content generator, content portions and presentation style which relate to the content associated with a basic logical object have been added. Of course, the value of the attribute object type is now basic logical object.

The description of the attributes can be found on p. 109 (application comments), 102 (bindings), 99 (content architecture class), 103 (content generator), 95 (content portions), 136 (enciphered), 97 (layout style), 88 (object class identifier), 88 (object type), 97 (presentation style), 137 (protection), 98 (resource), 136 (sealed) and 108 (user-readable comments and user-visible name).

It should be noted that the attributes enciphered and sealed were introduced by the Addendum on Security; they were not contained in the initially published version of the Standard.

3.1.4   The Constituents of the Generic Layout Structure

The constituents of the generic layout structure are the document layout root class, page set classes, page classes, frame classes and block classes.

Document layout root class

A document layout root class is represented by the following attribute set:

> *document layout root class* :=
>   {object class identifier, object type = document layout root,
>   [application comments], [bindings], [default value lists],
>   [generator for subordinates], [resource], [sealed],
>   [user-readable comments], [user-visible name]}

Similarly to the generic logical structures, there are some special rules for the attribute generator for subordinates which shall be recorded for later reference purposes:

Note 8:   The attribute generator for subordinates must be specified if the generic layout structure of a document is a complete generator set; it is not allowed if the generic layout structure is a factor set (see Sect. 3.2.15).

Note 9:   If the generic layout structure of a document is a partial generator set (see Sect. 3.2.15) the objects in the specific layout structure belonging to the object class may have subordinate objects which are not explicitly specified by the attribute generator for subordinates.

The description of the attributes can be found on p. 109 (application comments), 102 (bindings), 138 (default value lists), 93 (generator for subordinates), 88 (object class identifier), 88 (object type), 98 (resource), 136 (sealed) and 108 (user-readable comments and user-visible name).

It should be noted that the attribute sealed was introduced by the Addendum on Security; it was not contained in the initially published version of the Standard.

Page set class

A page set class is represented by the following attribute set:

page set class :=
  {object class identifier, object type = page set,
  [application comments], [bindings], [default value lists],
  [generator for subordinates], [resource], [sealed],
  [user-readable comments], [user-visible name]}

Concerning the attribute generator for subordinates, Notes 8 and 9 apply.

The description of the attributes can be found on p. 109 (application comments), 102 (bindings), 138 (default value lists), 93 (generator for subordinates), 88 (object class identifier), 88 (object type), 98 (resource), 136 (sealed) and 108 (user-readable comments and user-visible name).

It should be noted that the attribute sealed was introduced by the Addendum on Security; it was not contained in the initially published version of the Standard.

Page classes

As already explained above, ODA distinguishes in principle between two kinds of pages, namely composite pages and basic pages, and therefore there are composite page classes and basic page classes in the generic structures. Since the attribute sets for the two kinds of pages are quite different they shall be introduced as two different constituents. According to the specific structures, an ODA document can contain either composite page classes alone or basic page classes alone; a mixture of both is not permitted.

It should be noted that the value of the attribute object type is the same for both page classes, namely composite or basic page, i.e., the page classes cannot be distinguished by means of this attribute. While for the specific structures composite pages and basic pages can be distinguished by the presence of the attribute subordinates (see p. 50), this may be more complicated for composite page classes and basic page classes: The attribute **generator for subordinates** may be missing if the generic layout structure is a partial generator set or factor set.

In some cases the two page classes can be identified by the presence of a certain attribute – for instance, if the attribute **balance** is present it must be a composite page class, if the attribute **presentation attributes** is specified it must be basic page class – but in certain cases a distinction is only possible by means of other constituents of the document. If, for example, the generic layout structure contains frame classes or block classes, only composite page classes can appear in the document, or if a page class is referred to from a basic page it must be a basic page class.

Composite page class

A composite page class is represented by the following attribute set:

composite page class :=
  {object class identifier, object type = composite or basic page,
  [application comments], [balance], [bindings], [colour],
  [default value lists], [dimensions], [generator for subordinates],
  [medium type], [page position], [resource], [sealed], [transparency],
  [user-readable comments], [user-visible name]}

Concerning the attribute **generator for subordinates**, Notes 8 and 9 apply.

The attribute **balance** is only allowed if the immediate subordinate constituents of the page class are frame classes, not block classes. If it is specified the attribute **generator for subordinates** must also be present and the values of the two attributes must be consistent (see p. 125).

The description of the attributes can be found on p. 109 (**application comments**), 125 (**balance**), 102 (**bindings**), 134 (**colour**), 138 (**default value lists**), 119 (**dimensions**), 93 (**generator for subordinates**), 135 (**medium type**), 88 (**object class identifier**), 88 (**object type**), 134 (**page position**), 98 (**resource**), 136 (**sealed**), 133 (**transparency**) and 108 (**user-readable comments** and **user-visible name**).

It should be noted that the attribute **sealed** was introduced by the Addendum on Security; it was not contained in the initially published version of the Standard.

Basic page class

A basic page class is represented by the following attribute set:

basic page class :=
   {object class identifier, object type = composite or basic page,
   [application comments], [bindings], [colour],
   [content architecture class], [content generator], [content portions],
   [dimensions], [enciphered], [medium type], [page position],
   [presentation attributes], [presentation style], [resource], [sealed],
   [transparency], [user-readable comments], [user-visible name]}

Compared to a composite page class, the attributes balance, default value lists and generator for subordinates are missing and the attributes content architecture class, content generator, content portions, presentation attributes and presentation style, which are related to the actual content of a basic page, have been added.

The description of the attributes can be found on p. 109 (application comments), 102 (bindings), 134 (colour), 99 (content architecture class), 103 (content generator), 95 (content portions), 119 (dimensions), 136 (enciphered), 135 (medium type), 88 (object class identifier), 88 (object type), 134 (page position), 101 (presentation attributes), 97 (presentation style), 98 (resource), 136 (sealed), 133 (transparency) and 108 (user-readable comments and user-visible name).

It should be noted that the attributes enciphered and sealed were introduced by the Addendum on Security; they were not contained in the initially published version of the Standard.

Frame class

A frame class is represented by the following attribute set:

frame class :=
   {object class identifier, object type = frame,
   [application comments], [balance], [bindings], [border], [colour],
   [default value lists], [dimensions], [generator for subordinates],
   [layout path], [logical source], [permitted categories], [position],
   [resource], [sealed], [transparency], [user-readable comments],
   [user-visible name]}

Concerning the attribute generator for subordinates, Notes 8 and 9 are valid.

For the attribute balance the same restrictions as described for composite page classes apply: It is only allowed if the immediately subordinate constituents of the frame class are frame classes, not block classes. If it is specified the attribute generator for subordinates must also be present and the values of the two attributes must be consistent (see p. 125).

The attribute permitted categories (see p. 128) is only permitted for so-called *lowest level frame classes*, i.e., for frame classes whose immediately subordinate constituents are block classes.

The description of the attributes can be found on p. 109 (application comments), 125 (balance), 102 (bindings), 121 (border), 134 (colour), 138 (default value lists), 119 (dimensions), 93 (generator for subordinates), 110 (layout path), 106 (logical source), 88 (object class identifier), 88 (object type), 128 (permitted categories), 117 (position), 98 (resource), 136 (sealed), 133 (transparency) and 108 (user-readable comments and user-visible name).

It should be noted that the attribute sealed was introduced by the Addendum on Security; it was not contained in the initially published version of the Standard.

Block class

A block class is represented by the following attribute set:

block class :=
   {object class identifier, object type = block,
   [application comments], [bindings], [border], [colour],
   [content architecture class], [content generator], [content portions],
   [dimensions], [enciphered], [position], [presentation attributes],
   [presentation style], [resource], [sealed], [transparency],
   [user-readable comments], [user-visible name]}

The description of the attributes can be found on p. 109 (application comments), 102 (bindings), 121 (border), 134 (colour), 99 (content architecture class), 103 (content generator), 95 (content portions), 119 (dimensions), 136 (enciphered), 88 (object class identifier), 88 (object type), 117 (position), 101 (presentation attributes), 97 (presentation style), 98 (resource), 136 (sealed), 133 (transparency) and 108 (user-readable comments and user-visible name).

It should be noted that the attributes enciphered and sealed were introduced by the Addendum on Security; they were not contained in the initially published version of the Standard.

3.1.5 Content Portions and Styles

In this section three other types of constituents of an ODA document shall be introduced which can appear within all four kinds of structures, as the content portions, or which do not belong to any of the four structures, as the layout styles and presentation styles (see Fig. 14).

Content portion

A content portion is represented by the following attribute set:

> *content portion* :=
> { ⟨type of coding⟩,
>   [content identifier layout], [content identifier logical],
>   [alternative representation], [coding attributes],
>   [content information] }

The value of the attribute type of coding specifies in which way the actual content of the content portion (character, raster graphics or geometric graphics content) has been encoded. Since this depends on the content architecture the permitted values (and default values) of the attribute are not defined in Part 2 of the Standard but in Parts 6, 7 and 8 where the different content architectures are specified (see p. 101). The content architecture of the content portion is not defined at the content portion itself, but by the attribute content architecture class given for the objects with which a content portion is associated.

The attributes content identifier logical and content identifier layout have been classified as non-mandatory. This is correct when considering each of these attributes by itself, but at least one of the two attributes must be present, since a content portion must belong to the logical structure or the layout structure. (The simple-structured CCITT documents are again an exception: for these documents both attributes may be missing.) It is also possible that both attributes are present: then the content portion belongs to both structures, as is the case, for instance, for all content portions except for the company logo in the example of the business letter (see Fig. 3).

The actual content of the content portion, i.e., the character, raster graphics or geometric graphics content, is specified by the value of the attribute content information. It is here where (almost) everything that an author would regard as the content of his document finally appears in an ODA document. It may be rather surprising that this attribute is classified as non-mandatory, i.e., that it can be missing from a given content portion. In this case, the content portion does not contain any content and one could assume that the whole content portion would then be superfluous. However, besides the attribute content information there are some other ways to generate content in a document as will be described in Sect. 3.2.8. A content portion without the attribute content information can then supply attribute values which may be needed for the content generated with some other method.

The description of the attributes for the content portions can be found on p. 100 (alternative representation), 101 (coding attributes), 89 (content identifier layout), 89 (content identifier logical), 99 (content information) and 101 (type of coding).

It should be noted, however, that further attributes for content portions, depending on the content architecture, may appear. These are described in those parts of the Standard which specify the content architectures.

Layout style

A layout style is represented by the following attribute set:

layout style :=
  {layout style identifier,
    [block alignment], [concatenation], [derived from], [fill order],
    [indivisibility], [layout category], [layout object class],
    [new layout object], [offset], [same layout object], [sealed],
    [separation], [synchronization], [user-readable comments],
    [user-visible name]}

All constituents of the specific and generic logical structures can refer to a layout style by means of the attribute layout style. The attributes of a layout style basically represent information which guides the creation of the specific layout structure of a document from its specific logical and generic structures. The attributes of the layout styles are evaluated during the layout process (see Section 3.6.2).

The attribute layout style identifier must be present and represents the "name" of the layout style by which other constituent can refer to it. All the other attributes are non-mandatory; however, a layout style having only the attribute layout style identifier is of no use since it does not provide any information to the layout process.

The description of the attributes can be found on p. 113 (block alignment), 123 (concatenation), 97 (derived from), 112 (fill order), 130 (indivisibility), 128 (layout category), 124 (layout object class), 130 (new layout object), 114 (offset), 131 (same layout object), 136 (sealed), 115 (separation), 126 (synchronization) and 108 (user-readable comments and user-visible name).

It should be noted that the attribute derived from was included by the Addendum on Styles and the attribute sealed by the Addendum on Security; they were not present in the initially published version of the Standard.

Presentation style

A presentation style is represented by the following attribute set:

presentation style :=
   {presentation style identifier,
   [border], [colour], [derived from], [presentation attributes], [sealed],
   [transparency], [user-readable comments], [user-visible name]}

All constituents which are basic objects or basic object classes (i.e., basic logical objects, basic logical object classes, basic pages, basic page classes, blocks or block classes) can refer to a presentation style by means of the attribute presentation style identifier. The presentation styles basically represent information which guides the layout process and imaging process of a document.

The attribute presentation style identifier must be present, and represents the "name" of the presentation style by which other constituents can refer to it. All the other attributes are non-mandatory; however, a presentation style having only the attribute presentation style identifier is of no use since it does not provide any information to the layout or imaging process.

The description of the attributes can be found on p. 121 (border), 134 (colour), 97 (derived from), 101 (presentation attributes), 89 (presentation style identifier), 136 (sealed), 133 (transparency) and 108 (user-readable comments and user-visible name).

It should be noted that the attribute derived from was included by the Addendum on Styles and the attribute sealed by the Addendum on Security; they were not present in the initially published version of the Standard.

3.1.6   The Constituents of the Protected Document Part

The constituents of the protected document part – this concept was introduced by the Addendum on Security and not present in the initially published version of the Standard – are sealed document profile descriptions, enciphered document profile descriptions, pre-enciphered document body part descriptions and post-enciphered document body part descriptions (see Fig. 15).

Sealed document profile

A sealed document profile is represented by the following two attributes:

sealed document profile :=
{protected part identifier, sealed document profile information}

The description of the attributes can be found on p. 90 (protected part identifier), and 137 (sealed document profile information).

Enciphered document profile, post-enciphered document body part and pre-enciphered document body part

A enciphered document profile, post-enciphered document body part or pre-enciphered document body part is represented by the two attributes protected part identifier and enciphered information:

enciphered document profile :=
{protected part identifier, enciphered information}

post-enciphered document body part :=
{protected part identifier, enciphered information}

pre-enciphered document body part :=
{protected part identifier, enciphered information}

The description of the attributes can be found on p. 90 (protected part identifier) and 137 (enciphered information).

3.2   Attributes of the Document Structures

This section describes the attributes specified in Part 2 of the ODA Standard, especially their permitted values and their semantics.

3.2.1   Attributes for Specific and Generic Structures

As described in Sects. 3.1.1 – 3.1.5, most attributes can be used both in generic and specific structures; some attributes can appear only in generic or only in specific structures. As a general rule, the role of an attribute within a document depends whether it appears in the generic or in the specific structures:

- An attribute specified for a constituent of the specific structure defines a property of this constituent in a particular document.
- An attribute specified for a constituent of the generic structure, i.e., an attribute specified for an object class, usually defines a rule concerning a property of an object which belongs to this object class.

In many cases, however, the property of a particular object is not restricted to conform to the rule specified in the object class. In these cases, the attribute of the object class is used essentially for providing a default value for the respective attribute if the attribute is missing from the attribute set of a specific object.

(The attributes for content portions and styles are not considered in this section. As a rule of thumb, these attributes play a similar role as the attributes of the specific structures, if the content portion or style is referred from an object of the specific structure, and a similar role as the attributes of the generic structures, if the content portion or style is referred from an object class of the generic structure.)

Consider, for instance, the attribute subordinates which can only be specified for constituents of the specific structure. It specifies the objects which *are* immediately subordinate to a particular constituent (see p. 91). The comparable attribute in the generic structures is generator for subordinates which specifies a rule saying which objects *may be* immediately subordinate to an object belonging to the object class.

As an example for an attribute which can be specified for constituents of both the generic and the specific structures, consider the attribute colour. If this attribute is specified for a constituent of the specific structure (only pages, frames or blocks are possible) it specifies the colour of the area associated with the object.   If the attribute is specified for a constituent of the generic structure (a page class, frame class or block class) it has the role of a rule: the colour of the area for objects belonging to the object class is "usually" as specified for the object class. However, an object does not necessarily have to obey this rule but can ignore it by specifying the attribute colour explicitly for the object.

For several attributes the situation is even more complicated: some attributes can be specified for constituents of both the specific and generic structures but their permitted values may be different, depending on the kind of structure in which they appear. This will be explained in detail if such an attribute is introduced.

3.2.2   Data Types of Attribute Values

The values of the attributes are build according to certain rules, depending on the attribute, i.e., each attribute has an associated data type and the value of a certain attribute in a document is an instance of this data type.

The most elementary data types in the ODA Standard are *integers* $(\ldots, -2, -1, 0, 1, 2, \ldots)$, *non-negative integers* $(0, 1, 2, \ldots)$ and *positive integers* $(1, 2, \ldots)$.

The values of several attributes are *character strings*, usually of two different types: Firstly, character strings using characters from the *minimum subrepertoire* of ISO 6937, Part 2, (*"Coded character sets for text communication – Part 2: Latin alphabetic and non-alphabetic graphic characters"*), and, secondly, character strings whose character repertoire is defined in the document profile, usually by reference to ISO Standards. Since the character repertoires of these character strings are defined outside the ODA Standard the character strings shall also be considered elementary data types.

The values of some attributes can be arbitrary sequences of *octets* (sequences of eight bits) whose semantics is specified outside the Standard which shall also be regarded as elementary data types.

For later reference purposes the following four terms shall therefore be introduced:

– An *ISO-6937/2 character string* is sequence of characters from the minimum subrepertoire of ISO 6937, Part 2. These are the following characters: a...z (26 lower case letters), A...Z (26 upper case letters), 0...9 (10 digits) and ' ( ) , - . / : ? + = (11 special symbols).

- A *comments character string* is a sequence of characters from the character repertoire which is specified by the attribute comments character sets in the document profile.
- A *alternative representation character string* is a sequence of characters from the character repertoire which is specified by the attribute alternative representation character sets in the document profile.
- an *octet string* is a sequence of bits where the number of elements of the sequence is a multiple of eight.

Before specifying further data types, some additional notations shall be introduced:

$[\ldots]^+$ indicates that an entity can occur one or more times. (Example: $\{[a]^+\}$ is a set containing one or more elements a.)

$[\ldots]^*$ indicates that an entity can occur zero, one or more times.

$(\ldots \mid \ldots)$ denotes an alternative. (Example: $(a \mid b \mid c)$ is either a or b or c.)

$[\![\ldots]\!]$ denotes a sequence of entities. (Examples: $[\![a\ b\ c]\!]$ is a sequence consisting of the three elements a (first element), b (second element) and c (third element). $[\![[a]^+]\!]$ is a sequence consisting of one or more elements a, i.e., $[\![a]\!]$, $[\![a\ a]\!]$, $[\![a\ a\ a]\!]$ etc. It should be noted that compared to a set the elements of a sequence have an ordering, i.e., there is a first, second, third, etc., element.)

Several of the terms introduced below consist of more than one word, for instance, "*string function*". Such terms are sometimes enclosed in apostrophes ('*string function*') to avoid ambiguities, i.e., "*a b*" denotes two terms, one being "*a*" and the other one "*b*", whereas "'*a b*'" denotes one term with the name "*a b*". The apostrophes themselves have no meaning.

As was seen when introducing the constituents of ODA documents in the previous sections, there are several types of constituents, usually distinguished by the attribute object type. Of most of these types there exist usually several instances within a document. For example, a document will normally contain several constituents of type block in its specific layout structure.

To distinguish the constituents of the same type from each other, each constituent in a document has a "name". This name is specified by the attributes object identifier, object class identifier, layout style identifier, presentation style identifier, content identifier logical or content identifier layout, depending on the type of the constituent. The ODA Standard uses the following "naming conventions":

- The name of a constituent, i.e., the value of these six attributes, consists of a sequence of non-negative integers.
- The first number of the sequence is defined by the type of the constituent:
  "0" for constituents of the generic layout structure,
  "1" for constituents of the specific layout structure,
  "2" for constituents of the generic logical structure,
  "3" for constituents of the specific logical structure,
  "4" for layout styles and
  "5" for presentation styles.
  For content portions the first number is "0", "1", "2" or "3", depending on the structure in which the content portion appears.
- For the document layout root class, document layout root, document logical root class and document logical root, the sequence consists only of one element, i.e., the "name" of these types of constituents is $[\,0\,]$, $[\,1\,]$, $[\,2\,]$ or $[\,3\,]$, respectively.
- Within the specific structures, the sequences are built in such a way that each constituent "inherits" the sequence of its immediate superior constituent and adds one additional number to it. As a consequence, the number of elements in the sequence reflects the "distance" of the constituent from the document logical root or document layout root, respectively, and the tree structure of the constituents can be derived from these number sequences (see the example below).
- The content portions "inherit" the number sequence of the constituents with which they are associated, and add also one additional number.

For a better understanding consider again Fig. 3 of Sect. 2.1.1. Within an ODA document, the values of the attribute object identifier for the constituents representing the objects of the specific logical and layout structure might be as shown in Fig. 16.

Looking more closely at Fig. 16 several properties of the "naming conventions" can be identified. As already mentioned, the number of elements in the sequences reflects the hierarchical level of the respective constituent in the tree structure. The immediately superior object of a constituent can be identified by omitting the last number in the sequence, and the immediately subordinate objects can be found by looking for sequences which are identical except for one additional number at the end. In other words, the tree structure of the specific structures can be derived from the "names" of the constituents.

layout structure		logical structure	
object	object identifier	object	object identifier
letter layout	$[\![1]\!]$	letter	$[\![3]\!]$
letter area	$[\![1\ 0]\!]$	header	$[\![3\ 1]\!]$
header area	$[\![1\ 0\ 0]\!]$	body	$[\![3\ 2]\!]$
body area	$[\![1\ 0\ 3]\!]$	end	$[\![3\ 5]\!]$
end area	$[\![1\ 0\ 1]\!]$	address	$[\![3\ 1\ 0]\!]$
address area	$[\![1\ 0\ 0\ 0]\!]$	subject	$[\![3\ 1\ 1]\!]$
subject area	$[\![1\ 0\ 0\ 1]\!]$	date	$[\![3\ 1\ 2]\!]$
date area	$[\![1\ 0\ 0\ 2]\!]$	salutation	$[\![3\ 2\ 0]\!]$
logo area	$[\![1\ 0\ 0\ 3]\!]$	1st para	$[\![3\ 2\ 2]\!]$
salutation area	$[\![1\ 0\ 3\ 0]\!]$	2nd para	$[\![3\ 2\ 1]\!]$
1st para area	$[\![1\ 0\ 3\ 1]\!]$	3rd para	$[\![3\ 2\ 3]\!]$
2nd para area	$[\![1\ 0\ 3\ 2]\!]$	greetings	$[\![3\ 5\ 1]\!]$
$\vdots$	$\vdots$	$\vdots$	$\vdots$
name area	$[\![1\ 0\ 1\ 3]\!]$	name	$[\![3\ 5\ 3]\!]$
enclosure area	$[\![1\ 0\ 1\ 4]\!]$	enclosure	$[\![3\ 5\ 4]\!]$

Fig. 16: Possible values of the attribute object identifier for the objects of the business letter

It is not required that the sequences are build from consecutive numbers. For instance, the sequences $[\![1\ 0\ 2]\!]$, $[\![3\ 0]\!]$, $[\![3\ 3]\!]$ or $[\![3\ 4]\!]$ are missing in this example. Furthermore, the sequences need not necessarily reflect the ordering of the constituents in the document. For example, the "body area" has the sequence $[\![1\ 0\ 3]\!]$ and the "end area" the sequence $[\![1\ 0\ 1]\!]$ though the "body area" precedes the "end area" of the document. The ordering of the constituents is defined by the attribute subordinates (see p. 91).

Since a lot of attributes refer to other constituents using this naming convention several terms shall be introduced for later reference purposes:

-   A *layout-object-class-id* is a sequence of non-negative numbers, the first number being "0", i.e.:

    $$layout\text{-}object\text{-}class\text{-}id := [\![\,0\ [non\text{-}negative\ integer]^* \,]\!]$$

-   A *layout-object-id* is a sequence of non-negative numbers, the first number being "1", i.e.:

    $$layout\text{-}object\text{-}id := [\![\,1\ [non\text{-}negative\ integer]^* \,]\!]$$

- A *logical-object-class-id* is a sequence of non-negative numbers, the first number being "2", i.e.:

    $logical\text{-}object\text{-}class\text{-}id := [\![\, 2 \; [non\text{-}negative \; integer]^* \,]\!]$

- A *logical-object-id* is a sequence of non-negative numbers, the first number being "3", i.e.:

    $logical\text{-}object\text{-}id := [\![\, 3 \; [non\text{-}negative \; integer]^* \,]\!]$

Superior terms to these shall be defined:

- A *object-class-id* is either a *logical-object-class-id* or a *layout-object-class-id*, i.e.:

    $object\text{-}class\text{-}id := (logical\text{-}object\text{-}class\text{-}id \,|\, layout\text{-}object\text{-}class\text{-}id)$

- A *object-id* is either a *logical-object-id* or a *layout-object-id*, i.e.:

    $object\text{-}id := (logical\text{-}object\text{-}id \,|\, layout\text{-}object\text{-}id)$

Accordingly, the respective terms for layout styles, presentation styles and content portions are introduced:

- A *layout-style-id* is a sequence of two non-negative numbers, the first number being "4", i.e.:

    $layout\text{-}style\text{-}id := [\![\, 4 \; '\,non\text{-}negative \; integer\,' \,]\!]$

- A *presentation-style-id* is a sequence of two non-negative numbers, the first number being "5", i.e.:

    $presentation\text{-}style\text{-}id := [\![\, 5 \; '\,non\text{-}negative \; integer\,' \,]\!]$

- A *constituent identifier* shall denote the identifier of any of the preceding kinds of constituents, i.e.:

    $constituent \; identifier :=$
    $(object\text{-}class\text{-}id \,|\, object\text{-}id \,|\, layout\text{-}style\text{-}id \,|\, presentation\text{-}style\text{-}id)$

- A *content-portion-id* is a sequence of non-negative numbers with at least two elements, the first number being "1", "2", "3" or "4", i.e.:

    $content\text{-}portion\text{-}id := [\![\, (0 \,|\, 1 \,|\, 2 \,|\, 3) \; [non\text{-}negative \; integer]^+ \,]\!]$

The first number is determined by the kind of structure to which the content portion belongs: "0" if it belongs to the generic layout structure, "1" if it belongs to the specific layout structure, etc. If the last number is omitted the resulting sequence must be the *object-id* or *object-class-id* of a basic object or basic object class contained in the document.

As an example, consider Figs. 3 and 16. The *content-portion-id* of the content portions associated with the "address area", "logo area", "name area" and "name" could be, for instance, $[\![\,1\ 0\ 0\ 0\ 0\,]\!]$, $[\![\,1\ 0\ 0\ 3\ 0\,]\!]$, $[\![\,1\ 0\ 1\ 3\ 3\,]\!]$ or $[\![\,3\ 5\ 3\ 1\,]\!]$, respectively.

– A *protected-part-id* is a sequence of two non-negative numbers, the first number being "6,", "7", "8" or "9", i.e.:

$$protected\text{-}part\text{-}id := [\![\,(6\,|\,7\,|\,8\,|\,9)\ 'non\text{-}negative\ integer'\,]\!]$$

The first integer is "6" for a sealed document profile, "7" for an enciphered document profile, "8" for a pre-enciphered document body part and "9" for a post-enciphered document body part.

Within the values of several attributes, the names of so-called *layout categories* (see p. 128) are used. These names consist of character strings from the minimum subrepertoire of ISO 6937, Part 2. Such a name shall be called *layout-category-id*, i.e.:

$$layout\text{-}category\text{-}id := ISO\text{-}6937/2\ character\ string$$

3.2.3 Expressions as Attribute Values

The value of several attributes is an expression, i.e., the attribute has no fixed value but a "calculation rule" for the value is given. The evaluation of a specific value for the attribute is performed during the processing of the document, usually during the layout process.

As a simple example where expressions are needed, consider the following case: A document shall have a footline on each page containing the page number. The page numbers shall not be given by the author when editing the document but they should be generated automatically during the layout process. This could be achieved, for instance, by having in the generic layout structure a constituent of type page class with a subordinate block class for which the attribute **generator for subordinates** (see p. 93) is specified. The value of this attribute would be an expression whose evaluation during the layout process, i.e., when a page belonging to this page class is created, would generate the page number of the particular page.

The ODA Standard defines five types of expressions, namely *string expressions, numeric expressions, object identifier expressions, binding reference expressions* and *construction expressions*, which are described next.

String expressions

A string expression is either an *atomic string expression* or a sequence of two or more *atomic string expressions*. An *atomic string expression* is either a *string literal*, a *binding reference* or a *string function* applied to a numeric expression. The permitted *string functions* are MAKE-STRING, UPPER-ALPHA, LOWER-ALPHA, UPPER-ROMAN and LOWER-ROMAN, which can be summarized as:

> *string expression* :=
>   (*'atomic string expression'*
> | [[*'atomic string expression'* [*'atomic string expression'*]$^{+}$]])
>
> *atomic string expression* :=
>   (*'string literal'* | *'binding reference'*
> | *'string function'* *'numeric expression'*)
>
> *string function* :=
>   (MAKE-STRING | UPPER-ALPHA | LOWER-ALPHA
> | UPPER-ROMAN | LOWER-ROMAN)

(Some of the terms in these definitions are enclosed in apostrophes marking a term consisting of more than one word; see the corresponding explanation on p. 69.)

A *string literal* is an *octet-string* (see p. 69) whose interpretation is context dependent. For instance, an octet string is interpreted as a character string if it appears in the context of character content, as a raster image if it appears in the context of raster graphics content, and as a geometric picture if it appears in the context of geometric graphics content.

A *binding reference* is in principle the access to a "variable" which has been declared by means of the attribute bindings. Binding references are explained in detail on p. 80; see also the description of the attribute bindings on p. 102. The application of a string function to a numeric expression – the numeric expression is evaluated *before* the string function is applied – creates a character or a character string according to the following rules:

– MAKE-STRING applied to a number produces a character string representing the number as a sequence of decimal digits.

– UPPER-ROMAN and LOWER-ROMAN applied to a number produce a character string representing the number as a sequence of Roman numerals, either with upper-case or lower-case letters.

- UPPER-ALPHA and LOWER-ALPHA applied to a number produce a character string consisting of one of the 26 letters "A" to "Z" or "a" to "z", respectively, corresponding to the value of the number which should be between 1 and 26. (Otherwise, an empty character string is generated.)

Some examples for clarification: "MAKE-STRING 13" produces "13". (The first 13 is a number, which might be the result of the evaluation of a numeric expression, the second 13 is a character string consisting of the two characters "1" and "3".) "UPPER-ROMAN 13" produces the character string "XIII" and "LOWER-ROMAN 13" generates "xiii" accordingly. The result of "UPPER-ALPHA 13" is "M" and of "LOWER-ALPHA 13" "m" since this is the thirteenth letter of the alphabet.

If a string expression consists of sequence of *atomic string expressions* each of them is evaluated separately and the resulting character strings are concatenated to a single character string. Therefore, the evaluation of a string expression always produces one (possibly empty) character string.

Numeric expressions

A *numeric expression* is either an integer, a *binding reference* or one of the *numeric functions* INCREMENT, DECREMENT or ORDINAL, applied to an argument. For the numeric functions INCREMENT and DECREMENT the argument is a numeric expression again, for the function ORDINAL it is either an *object-id* or an object identifier expression. This can be summarized as:

```
numeric expression :=
    (integer | 'binding reference'
    | INCREMENT 'numeric expression'
    | DECREMENT 'numeric expression'
    | ORDINAL object-id
    | ORDINAL 'object identifier expression')
```

The result of the functions INCREMENT and DECREMENT is a number which is one greater or one less than the value of the numeric expression to which the functions are applied.

The result of the function ORDINAL is the sequence number of the object – either specified directly by an *object-id* or indirectly by an object identifier expression – within the set of its neighboring objects according to their so-called *sequential order* (see p. 91). Objects are called neighboring if they have the same immediate superior object and belong to the same object class.

Consider Fig. 21 for two examples:

− The value of the numeric expression "ORDINAL [[3 1 2]]" is "3" because the neighboring objects are those with the *object-ids* [[3 1 0]] and [[3 1 1]] and according to the sequential order the object with the *object-id* [[3 1 2]] is the third.

− If the value of an attribute for the object with the *object-id* [[3 2 2]] is "ORDINAL CURRENT-OBJECT" (the function CURRENT-OBJECT is explained below) the evaluation of this numeric expression yields the value "2" because there is only one preceding object according to the sequential order, the one with the *object-id* [[3 2 0]].

Object identifier expressions

An *object identifier expression* consists of one of the four *object selection functions* CURRENT-OBJECT, SUPERIOR-OBJECT, PRECEDING-OBJECT or CURRENT-INSTANCE which have zero, one or two arguments:

> *object identifier expression* :=
>   (CURRENT-OBJECT
> | SUPERIOR-OBJECT ' *object identifier expression* '
> | PRECEDING-OBJECT ' *object identifier expression* '
> | CURRENT-INSTANCE (' *object class identifier* ' | ' *object type* ')
>   (' *object identifier expression* ' | *object-id*))

The evaluation of an object identifier expression always yields an *object-id* (see p. 72). The function CURRENT-OBJECT creates the *object-id* of the object which caused the evaluation of the object identifier expression.

An example for clarification: An object of a specific structure may have an attribute whose value is defaulted from the object class to which the object belongs, and the value of this attribute for the object class may be the object identifier expression "CURRENT-OBJECT". When the value of the attribute for the constituent of the specific structure is required, for instance, at some stage during the layout process, it will then become the *object-id* of the object. In other words, the function CURRENT-OBJECT provides a means for an object to refer to its own "name", usually when it is needed in another expression.

The function **SUPERIOR-OBJECT** creates the *object-id* of the object which is immediately superior to the object specified by the argument of the function. The argument of the function, the object identifier expression, is evaluated *before* the function is applied, i.e., the function is actually applied always to an *object-id*.

Two examples for clarification:

- The evaluation of the object identifier expression "SUPERIOR-OB-JECT ⟦1 0 3⟧" – it shall be assumed that the object identifier expression which was the argument of function has evaluated to ⟦1 0 3⟧ – yields the *object-id* ⟦1 0⟧.
- The evaluation of the object identifier expression "SUPERIOR-OB-JECT CURRENT-OBJECT" yields the *object-id* of the object which is immediate superior to the one which caused the evaluation of the object identifier expression.

The function **PRECEDING-OBJECT** creates the *object-id* of the object which immediately precedes the object specified by the argument of the function according to their sequential order. Again, the argument of the function, the object identifier expression, is evaluated *before* the function is applied.

Consider Fig. 21 for two examples:

- The value of the object identifier expression "PRECEDING-OBJECT ⟦3 2 2⟧" – it shall be assumed that ⟦3 2 2⟧ was the result of the evaluation of the argument of the function – is ⟦3 2 0⟧.
- If the object identifier expression "PRECEDING-OBJECT SUPERIOR-OBJECT CURRENT-OBJECT" is evaluated for the object with the *object-id* ⟦3 5 4⟧ it yields ⟦3 2⟧.

The function **CURRENT-INSTANCE** has two arguments. The first argument is either an *object-class-id* or an *object type*, i.e., one of the possible values of the attribute **object type** (see p. 88). The second argument is either an object identifier expression or an *object-id*. The result is always an *object-id* which is determined according to the following four methods:

1. The first argument specifies a logical object class or a logical object type (**document logical root**, **composite logical object** or **basic logical object**), and the second argument refers to an object of the specific logical structure which shall be called the *reference object*.
   - If the reference object belongs to the object class specified by the first argument or is of the object type indicated by the first

argument, the result of the function is the *object-id* of the reference
object.

— Otherwise, the result of the function is the *object-id* of that object
which belongs to the object class specified by the first argument
or is of the object type indicated by the first argument and is
"nearest" to the reference object. The search for the "nearest"
object is performed in ascending order within the specific logical
structure, i.e., in the direction of the document logical root.

An example for clarification: ⟦3 4 1⟧ shall be the *object-id* of a
basic logical object. The result of

CURRENT-INSTANCE 'composite logical object' ⟦3 4 1⟧

is then the *object-id* of the composite logical object to which the basic
logical object is immediately subordinate, i.e., ⟦3 4⟧.

2. The first argument specifies a layout object class or a layout object
type (document layout root, page set, composite or basic page, frame
or block), and the second argument refers to an object of the specific
logical structure or a temporary logical object which has been created
as a result of the attribute logical source (see p. 106). In this case,
the *reference object* is that basic layout object in which the content of
the logical object is laid out. (If the layout process splits the content
into more than one basic layout object the first one is the reference
object.)

— If the reference object belongs to the object class specified by
the first argument or is of the object type indicated by the first
argument, the result of the function is the *object-id* of the reference
object.

— Otherwise, the result of the function is the *object-id* of that object
which belongs to the object class specified by the first argument
or is of the object type indicated by the first argument and is
"nearest" to the reference object. The search for the "nearest"
object is performed in ascending order within the specific layout
structure, i.e., in the direction of the document layout root.

Example: ⟦3 4 1⟧ shall be the *object-id* of a basic logical object.
Then the result of

CURRENT-INSTANCE 'composite or basic page' ⟦3 4 1⟧

is the *object-id* of the page on which the content associated with the
basic logical object is laid out.

3. The first argument specifies a logical object class or a logical object
type (document logical root, composite logical object or basic logical
object), and the second argument refers to an object of the specific

layout structure whose object class is referred from at least one basic layout object without generic content (content that is created by a content portion of the generic structures). In this case, the *reference object* is the first logical object whose associated content is laid out in the layout object specified by the second argument.

– If the reference object belongs to the object class specified by the first argument or is of the object type indicated by the first argument, the result of the function is the *object-id* of the reference object.

– Otherwise, the result of the function is the *object-id* of that object which belongs to the object class specified by the first argument or is of the object type indicated by the first argument and is "nearest" to the reference object. The search for the "nearest" object is performed in ascending order within the specific logical structure, i.e., in the direction of the document logical root.

Example: $[\![1\ 2\ 5]\!]$ shall be the *object-id* of a page. The result of

CURRENT-INSTANCE 'basic logical object' $[\![1\ 2\ 5]\!]$

is the *object-id* of that basic logical object which contributes as the first one to the actual content of the page.

4. The first argument specifies a logical object class or a logical object type (**document logical root, composite logical object** or **basic logical object**), and the second argument refers to a temporary logical object which has been created as a result of the attribute **logical source** (see p. 106). In this case, a *reference layout object* and a *reference logical object* are defined.

The *reference layout object* is that basic layout object into which the content associated with the temporary logical object is laid out. (If the layout process splits the content into more than one basic layout object the first one is the reference layout object.) Then the basic layout object is determined which follows the reference layout object in sequential order and which receives content from one or more basic logical objects (not temporary logical objects). The first of these basic logical objects is called the *reference logical object*.

– If the reference logical object belongs to the object class specified by the first argument or is of the object type indicated by the first argument, the result of the function is the *object-id* of the reference logical object.

– Otherwise, the result of the function is the *object-id* of that object which belongs to the object class specified by the first argument or is of the object type indicated by the first argument and is "nearest" to the reference logical object. The search for the "nearest" object is performed in ascending order within the specific logical structure, i.e., in the direction of the document logical root.

If an *object-id* cannot be determined according to one of these four methods the result of the function is an empty *object-id* (⟦ ⟧). This would happen, for instance, when evaluating

CURRENT-INSTANCE 'page set' ⟦3 4 1 0⟧

for a document containing no page sets.

Binding references and binding reference expressions

A preliminary remark for readers with a background in programming languages: A *binding reference* can be viewed as an access to a "variable" which is stored at some place in the document. The ODA Standard does not use the term "variable" but, in principle, the concept of variables is introduced with the attribute bindings (see p. 102), however, with a rather limited functionality.

A *binding reference* consists of a *binding reference expression* followed by a *binding name*. A *binding reference expression* is either an *object-id* or one of the four *binding selection functions* CURRENT-OBJECT, SUPE-RIOR, PRECEDING or CURRENT-INSTANCE, which have no argument (CURRENT-OBJECT), one argument (SUPERIOR and PRECEDING) or two arguments (CURRENT-INSTANCE), i.e.:

> *binding reference* :=
> ⟦ '*binding reference expression*' '*binding name*' ⟧

> *binding reference expression* :=
>   (*object-id* | CURRENT-OBJECT
> | SUPERIOR '*object identifier expression*'
> | PRECEDING '*object identifier expression*'
> | CURRENT-INSTANCE ('*object class identifier*' | '*object type*')
>   ('*object identifier expression*' | *object-id*))

A *binding name* is a character string from the minimum subrepertoire of ISO 6937, Part 2. Thinking in terms of variables, the binding name is essentially the name of the variable and the binding reference expression denotes the place in the document where the variable is stored.

The two binding selection functions CURRENT-INSTANCE and CUR-RENT-OBJECT have the same semantics and same arguments as the corresponding object selection functions introduced above.

The semantics of the binding selection functions SUPERIOR and PRE-CEDING are similar to the object selection functions SUPERIOR-OBJECT and PRECEDING-OBJECT with the following differences.

If the attribute bindings is not specified for the object whose *object-id* is the result of the object identifier expression or – in the case that the attribute is specified – if the binding name, given for the binding reference, cannot be found in the value of this attribute, a search for another object with this binding name is started. The direction in which the search is performed depends on the binding selection function.

For the function SUPERIOR, the search is performed in the direction of the document logical root or document layout root, i.e., if the binding name cannot be found on an object the search continues at its immediately superior object until either the binding name is found or the document logical root or document layout root, respectively, is reached.

For the function PRECEDING, the search is performed in the direction opposite to the sequential order of the objects, i.e., if the binding name cannot be found on an object, the search continues at the object which precedes in the sequential order until either the binding name is found or the document logical root or document layout root, respectively, is reached.

If the binding name is found, the result of the binding reference is the value associated with the binding name (see p. 102). This value may be an expression again which will then be evaluated. Ultimately, however, the binding reference will always resolve to a character string, an integer or an *object-id*, depending on the context. If the binding name cannot be found anywhere its value is an empty character string, the integer 0 or the empty *object-id* [[ ]], again depending on the context in which the binding reference is used.

Consider Fig. 21 for two examples:

- If the binding reference "SUPERIOR CURRENT-OBJECT abc" shall be evaluated for the object with the *object-id* [[3 2 2]] it is first examined whether the attribute bindings is specified for the object with the *object-id* [[3 2]] (or for the object class to which this object may belong) and whether the binding name abc is present in the value of the attribute bindings. If yes, the value associated with this binding name is the result of the binding reference. Otherwise an additional attempt to find a value for the binding reference is made with the document logical root, the object with the *object-id* [[3]].

- If the binding reference "PRECEDING CURRENT-OBJECT abc" shall be evaluated for the object with the *object-id* [[3 2 2]] the first attempt to find the binding name is made with the object which has the sequence number 7 in the sequential order, in case of failure with the object with sequence number 6, then with the number 5, etc., until the document logical root is reached.

Construction expressions

The values of the attribute **generator for subordinates** (see p. 93) are so-called *construction expressions* which are defined recursively in the ODA Standard: A construction expression is either a *construction type* or a *construction term*. A construction type is a construction term which has as a prefix either **sequence**, **aggregate** or **choice**. A construction term is either an *object-class-id* or construction type, prefixed with either **req** (required), **opt** (optional), **rep** (repetitive) or **optrep** (optional repetitive). This can be summarized as:

construction expression :=
  (' *construction type* ' | ' *construction term* ')

construction type :=
  (**sequence** [' *construction term* ']$^+$
  | **aggregate** [' *construction term* ']$^+$
  | **choice** [' *construction term* ']$^+$)

construction term :=
  (**req** (*object-class-id* | ' *construction type* ')
  | **opt** (*object-class-id* | ' *construction type* ')
  | **rep** (*object-class-id* | ' *construction type* ')
  | **optrep** (*object-class-id* | ' *construction type* '))

Since the concept of construction expressions is somewhat complicated because of the recursive definition it shall be explained by some examples, but first the meaning of the prefixes shall be specified.

The prefix **sequence** specifies that the elements of the construction term which follows are ordered as a sequence, i.e., there is a first, second, third, etc., element.

The prefix **aggregate** specifies that the elements of the construction term which follows are unordered, i.e., they build a set.

The prefix **choice** specifies that the elements of the construction term are considered alternatives, i.e., when evaluating the construction expression, exactly one element must be selected.

The prefix **req** specifies that the term which follows (an *object-class-id* or a construction type) must be present when evaluating the construction expression, and **opt** specifies that the term is optional, i.e., it need not necessarily be present.

The prefix **rep** specifies that the term which follows is present one or more times, and **optrep** specifies that it can be present zero, one or more times.

The first step when evaluating a construction expression is a recursive resolution of the terms until only the prefixes and *object-class-ids* remain. Resolving the result further by considering the semantics of the prefixes will finally lead to a (possibly empty) sequence of *object-class-ids*.

For clarification some examples based on the generic logical structure of the business letter as shown in Fig. 4 will be given. Firstly, *object-class-ids* shall be assigned to some of the object classes of this generic structure which, for instance, may be the ones shown in Fig. 17.

object class	object-class-id	object class	object-class-id
letter	[2]	body content	[2 2 1]
header	[2 0]	text	[2 2 1 0]
body	[2 2]	figure	[2 2 1 1]
end	[2 1]	greetings	[2 1 0]
address	[2 0 1]	signature	[2 1 2]
date	[2 0 2]	name	[2 1 1]
subject	[2 0 3]	enclosure	[2 1 3]

Fig. 17: Values of the attribute **object class identifier** for object classes of the business letter

In this case, the values of the attribute **generator for subordinates** for some of the object classes would be as shown in Fig. 18 (see also Fig. 4 and the explanation of the graphical notation which is used there).

To understand that these attribute values are really construction expressions, consider the following value for which its decomposition into the terms used in the definition of construction expressions is shown:

sequence req [2 0 1] req [2 0 2] opt [2 0 3]

object-class-id   *object-class-id*   *object-class-id*

construction term   *construction term*   *construction term*

construction type

object-class-id	value of the attribute generator for subordinates
[2]	sequence req [2 0] req [2 2] req [2 1]
[2 0]	sequence req [2 0 1] req [2 0 2] opt [2 0 3]
[2 1]	sequence req [2 1 0] req [2 1 2] req [2 1 1] opt [2 1 3]
[2 2]	sequence req [2 2 1] optrep [2 2 1]
[2 2 1]	choice [2 2 1 0] [2 2 1 1]

Fig. 18: Values of the attribute generator for subordinates for object classes of the business letter

(The prefix req is usually omitted in ODA documents; it is explicitly specified here for clarification.)

The following difference from the four kinds of expressions explained above should be noted: While the evaluation of these expressions always yields exactly one result (a character string for a string expression, an integer for a numeric expression, an *object-id* for an object identifier expression or a binding value for a binding reference expression) the evaluation of a construction expression may create a set of values.

For instance, consider once again the construction expression "sequence req [2 0 1] [2 0 2] opt [2 0 3]". Its evaluation may yield both "[[2 0 1] [2 0 2]]" and "[[2 0 1] [2 0 2] [2 0 3]]" as a result. The evaluation of "sequence req [2 2 1] optrep [2 2 1]" may even yield an infinite number of results, "[[2 2 1]]", "[[2 2 1] [2 2 1]]", "[[2 2 1] [2 2 1] [2 2 1]]", etc., which are in the value range of this construction expression. This will be discussed in greater detail with the attribute generator for subordinates (see p. 93).

3.2.4   Classification of the Attributes

In Sects. 3.1.1 – 3.1.5 all attributes for describing the structures of ODA documents had already appeared when the constituents of the documents were explained, but the attributes, especially their values, had not been discussed in detail. This is done in Sects. 3.2.5 – 3.2.13.

As could be seen already, there are attributes which can be specified for constituents of both the logical structure and the layout structure. These attributes are called *shared attributes*. Several attributes can only be specified for either constituents of the logical or the layout structure. Such attributes are called *logical attributes* or *layout attributes*, respectively. Furthermore, some attributes can be specified only for layout styles, presentation styles or content portions. Accordingly, they are called *layout style attributes, presentation style attributes* or *content portion attributes*.

Figure 19 lists all attributes described in Part 2 of the Standard and indicates for which types of constituents they can be specified. (The attributes for constituents of the protected part of a document are not included in this list. These are the attributes **enciphered information, protected part identifier** and **sealed document profile information**.) The table is read as follows: A "–" in a table entry indicates that the respective attribute cannot be used for the types of constituent given in the header of the table. The other table entries consist usually of two letters, where "m" stands for mandatory, "d" for defaultable and "n" for non-mandatory, whose interpretation depends on the column in which the letters appear. (Again, "–" indicates that the attribute cannot be used.)

In the first two columns (logical structure and layout structure), the first letter indicates whether the respective attribute is mandatory or non-mandatory in the generic structures, the second letter (after the "/") indicates whether the respective attribute is mandatory, defaultable or non-mandatory in the specific structures. In several cases, however, the classification does not apply to all constituents of the respective structures but only to a subset of them, or additional constraints may be valid. This is indicated in the table by superscripts referring to footnotes in the table.

For content portions, the first letter indicates the classification of the attribute if the content portion appears in the generic structures, the second if it appears in the specific structures.

In the last two columns (layout styles and presentation styles), the first letter indicates whether the attribute is mandatory or non-mandatory if the style is referenced from an object class of the generic structures, the second whether it is mandatory, defaultable or non-mandatory if the style is referenced from an object of the specific structures.

It should be noted that the role of an attribute specified for a style is slightly different from the role of attributes specified for the other types of constituents: An attribute specified for an object, an object class or a content portion describes a property of the constituent for which it is specified. An attribute specified for a style describes a property of the constituents of the logical or layout structures which refer to this style by means of the attribute **layout style** or **presentation style**. The attributes specified for styles are applied to the constituents which refer to them during the layout process or imaging process.

attribute	logical structure	layout structure	content portions	layout styles	presenta-tion styles
alternative	$-/n^{22}$	$-/n^{22}$	–	–	–
alternative representation	–	–	n/n	–	–
application comments	n/d	n/d	–	–	–
balance	–	n/d^1	–	–	–
bindings	n/d	n/d	–	–	–
block alignment	–	–	–	n/d^2	–
border	–	n/d^3	–	–	n/d
coding attributes	–	–	n/n^4	–	–
colour	–	n/d^5	–	–	n/d
concatenation	–	–	–	n/d^2	–
content architecture class	n/d^6	n/d^6	–	–	–
content generator	n/n^7	$n/-^7$	–	–	–
content identifier layout	–	–	n/n	–	–
content identifier logical	–	–	n/n	–	–
content information	–	–	n/n	–	–
content portions	$n/n^{6,16}$	$n/n^{6,16}$	–	–	–
default value lists	n/n^8	n/n^8	–	–	–
derived from	–	–	–	n/n	n/n
dimensions	–	n/d^5	–	–	–
enciphered	n/n^{23}	n/n^{23}	–	–	–
fill order	–	–	–	n/d^2	–
generator for subordinates	$n/-^9$	$n/-^9$	–	–	–
imaging order	–	$-/n^{10}$	–	–	–
indivisibility	–	–	–	n/d^{11}	–
layout category	–	–	–	n/d^2	–
layout object class	–	–	–	n/d	–
layout path	–	n/d^{12}	–	–	–
layout style identifier	–	–	–	m/m	–
layout style	n/n	–	–	–	–
logical source	–	$n/-^{13}$	–	–	–
medium type	–	n/d^{14}	–	–	–
new layout object	–	–	–	n/d^{11}	–
object class identifier	m/-	m/-	–	–	–
object class	$-/n^{15}$	$-/n^{15}$	–	–	–
object identifier	$-/m^{16}$	$-/m^{16}$	–	–	–
object type	m/d^{17}	m/d^{17}	–	–	–
offset	–	–	–	n/d^2	–
page position	–	n/d^{14}	–	–	–

Fig. 19: Relations between attributes and constituents

attribute	logical structure	layout structure	content portions	layout styles	presentation styles
permitted categories	–	n/d^{18}	–	–	–
position	–	n/d^3	–	–	–
presentation attributes	–	n/d^{19}	–	–	n/d
presentation style identifier	–	–	–	–	m/m
presentation style	n/n^6	n/n^6	–	–	–
primary	$-/n^{21,22}$	$-/n^{21,22}$	–	–	–
protection	n/d	–	–	–	–
resource	n/–	n/–	–	–	–
same layout object	–	–	–	n/d^{11}	–
sealed	n/d	n/d	–	n/n	n/n
separation	–	–	–	n/d^2	–
subordinates	$-/m^{16,20}$	$-/m^{16,20}$	–	–	–
synchronization	–	–	–	n/d^{11}	–
transparency	–	n/d^5	–	–	n/d
type of coding	–	–	d/d	–	–
user-readable comments	n/d	n/d	–	n/n	n/n
user-visible name	n/d	n/d	–	n/n	n/n

1: only permitted for composite objects and composite object classes without immediately subordinate blocks;   2: only applicable for basic logical objects and basic logical object classes;   3: only permitted for frames, blocks, frame classes and block classes;   4: depending on content architecture;   5: only permitted for pages, frames, blocks, page classes, frame classes and block classes;   6: only permitted for basic objects and basic object classes;   7: only permitted for basic logical objects and basic object classes;   8: only permitted for composite objects and composite object classes;   9: only permitted for composite object classes;   10: only permitted for composite pages and frames;   11: not applicable for document logical roots and document logical root classes;   12: only permitted for frames and frame classes;   13: only permitted for frame classes;   14: only permitted for pages and page classes;   15: mandatory for logical objects and composite layout objects in the case of complete generator sets; not permitted in the case of factor sets;   16: not mandatory for simple-structured CCITT documents;   17: mandatory for constituents of the specific structure if the attribute **object class** is not specified;   18: only permitted for lowest level frames and lowest level frame classes;   19: only permitted for basic pages, blocks, basic page classes and block classes;   20: only permitted for composite objects;   21: mandatory for alternative representations;   22: only permitted for basic objects;   23: not permitted for composite object classes

Fig. 19: (continued)

3.2.5   Attributes for the Identification of Constituents

Within the attribute set of the constituents there appear seven attributes which are used for the identification of the constituents, namely object type, object identifier, object class identifier, layout style identifier, presentation style identifier, content identifier logical and content identifier layout.

The possible values of the attribute object type were already given when the constituents were introduced above. This can be summarized as:

object type =
  (document logical root | composite logical object |
    basic logical object | document layout root | page set |
    composite or basic page | frame | block)

This attribute must be specified for all object classes. It can be missing for constituents of the specific structures but then the attribute object class must be specified. In this case, the value of the attribute object type of the referenced object class determines the type of the object. In other words, each constituent in the generic or specific, logical or layout structures of a document has a certain type which is either determined directly by the the attribute object type or indirectly by a reference to an object class.

The value of the attribute object identifier is either a *layout-object-id* or *logical-object-id*, and the value of object class identifier is a *layout-object-class-id* or *logical-object-class-id* (see pp. 69 – 73 for the definition of these terms), i.e.:

object identifier = (*layout-object-id* | *logical-object-id*)

and

object class identifier = (*layout-object-class-id* | *logical-object-class-id*)

The attribute object identifier or object class identifier, respectively, must always be specified. (Exceptions to this rule are the so-called *simple structured CCITT documents*, see Sect. 3.3.3.) The values of these attributes must be unique within a document, i.e., no constituents are permitted within a document for which the value of the attribute object identifier or object class identifier consists of the same sequence of numbers. This uniqueness requirement, which also holds for the attributes

layout style identifier, presentation style identifier, content identifier logical and content identifier layout described below, should be obvious: the
identifiers are the "names" of the constituents and must therefore be
chosen in such a way that the constituents can be uniquely identified by
them.

The value of the attribute layout style identifier is a *layout-style-id* (see
p. 72):

layout style identifier = *layout-style-id*

This attribute must be specified for layout styles and represents the
"name" by which a layout style can be referenced from objects or object
classes of the logical structure by means of the attribute layout style.

The value of the attribute presentation style identifier is a *presentation-
style-id* (see p. 72):

presentation style identifier = *presentation-style-id*

This attribute must be specified for presentation styles and represents
the "name" by which a presentation style can be referenced from objects
or object classes by means of the attribute presentation style identifier.

The values of the attributes content identifier logical and content identifier layout are *content-portion-ids*:

content identifier logical = *content-portion-id*

content identifier layout = *content-portion-id*

These attributes can be specified for content portions and represent
the "names" by which the content portions can be referenced from basic
logical object classes, basic layout object classes, basic logical objects
and basic layout objects by means of the attribute content portions (see
p. 95). At least one of the attributes must be specified, depending on
the structure to which the content portion belongs. Both attributes can
be present if a content portion belongs to the logical and also the layout
structure.

The value of the attribute alternative is either a *layout-object-id* or
logical-object-id:

alternative = (*layout-object-id* | *logical-object-id*)

This attribute can be specified for basic objects (basic logical objects, basic pages and blocks) and indicates the "name" of another basic object which serves as an alternative description of the object for which the attribute is specified (see Sect. 3.2.14), if such an alternative description exists in the document. If the object is a basic logical object, the value of the attribute is a *logical-object-id*; if it is a basic layout object, the value of the attribute is a *layout-object-id*.

The attribute alternative – and the attribute primary described below – were introduced by the Addendum on Alternate Descriptions. They were not present in the initially published version of the Standard.

The value of the attribute primary is either a *layout-object-id* or *logical-object-id*:

primary $= (layout\text{-}object\text{-}id \mid logical\text{-}object\text{-}id)$

This attribute can be specified for basic objects (basic logical objects, basic pages and blocks) and indicates the "name" of another basic object which is the primary description of the object for which the attribute is specified (see Sect. 3.2.14). If the object is a basic logical object, the value of the attribute is a *logical-object-id*; if it is a basic layout object, the value of the attribute is a *layout-object-id*.

For an object to be an alternative description, this attribute must always be specified, i.e., the presence of this attribute for a basic object identifies the object as an alternative description.

The value of the attribute protected part identifier is a *protected-part-id*:

protected part identifier $= protected\text{-}part\text{-}id$

This attribute can be specified for constituents of the protected part and represents the "name" by which these constituents can be referenced from constituents of the unprotected parts by means of the parameter protected part identifier of the attribute enciphered and also by several parameters of document profile attributes related to security (see Sect. 4.2.12).

3.2.6   Attributes for the Creation of Structures

Three attributes are used to create the structures of ODA documents, namely subordinates, generator for subordinates and content portions. The hierarchical (tree) structures in the specific structures are described by the attribute subordinates. The corresponding attribute in the generic structures is generator for subordinates which specifies the rules for the structures. The attribute content portions associates the content portions in a document with constituents in the logical and layout structures.

The value of the attribute subordinates is a sequence of non-negative numbers, i.e.:

$$\textsf{subordinates} = [\![\,[non\text{-}negative\ integer]^{+}\,]\!]$$

This attribute must be specified for all composite objects of the logical structure and layout structure, i.e., for the document logical root and for composite logical objects, for the document layout root, for page sets, composite pages and frames. (Exceptions are again the simple-structured CCITT documents, for which the attribute can be missing.)

The attribute describes the objects which are immediately subordinate to the object for which it is specified. More precisely: if the value of the attribute object identifier for a particular object is $[\![\,x_1\ x_2\ \dots\ x_n\,]\!]$ and the value of the attribute subordinates is $[\![\,y_1\ y_2\ \dots\ y_m\,]\!]$, the immediately subordinate objects are those for which the value of the attribute object identifier is $[\![\,x_1\ x_2\ \dots\ x_n\ y_1\,]\!]$, $[\![\,x_1\ x_2\ \dots\ x_n\ y_2\,]\!]$ $\dots$ $[\![\,x_1\ x_2\ \dots\ x_n\ y_m\,]\!]$. Within the sequence $[\![\,y_1\ y_2\ \dots\ y_m\,]\!]$ no number may appear twice, which should be rather obvious.

The ordering of the numbers within the sequence of the attribute value of subordinates is of importance: it defines the so-called *sequential order* of the objects in the logical structure (*sequential logical order*) and in the layout structure (*sequential layout order*). This sequential order is of relevance for the layout process and imaging process.

The sequential logical order determines the order in which the objects of the logical structure shall be processed by the layout process. The sequential layout order determines the order in which the objects of the layout structure shall be processed by the imaging process.

If, for instance, two layout objects are (totally or partially) laid out in the same area on the presentation medium, the appearance of the document on paper or on a computer screen may depend on which of the two layout objects is imaged first. (The order in which layout objects are imaged on the presentation medium can also be influenced by the attribute imaging order which must be specified if the order shall be different from the sequential layout order; see p. 132.)

For clarification consider again Figs. 3 and 16. The possible values of the attribute subordinates could then be as shown in Fig. 20.

layout structure		logical structure	
object	subordinates	object	subordinates
letter layout	$[\![\,0\,]\!]$	letter	$[\![\,1\ 2\ 5\,]\!]$
letter area	$[\![\,0\ 3\ 1\,]\!]$	header	$[\![\,0\ 1\ 2\,]\!]$
header area	$[\![\,0\ 1\ 2\ 3\,]\!]$	body	$[\![\,0\ 2\ 1\ 3\,]\!]$
body area	$[\![\,0\ 1\ 2\ 3\,]\!]$	end	$[\![\,1\ 2\ 3\ 4\,]\!]$
end area	$[\![\,1\ 2\ 3\ 4\,]\!]$		

Fig. 20: Possible values of the attribute subordinates for the objects of the business letter

A value of $[\![\,0\ 1\ 2\ 3\,]\!]$ for the attribute subordinates, as specified for the layout object "header area", for instance, specifies that the immediately subordinate objects of the "header area" are those with the *object-ids* $[\![\,1\ 0\ 0\ 0\,]\!]$, $[\![\,1\ 0\ 0\ 1\,]\!]$, $[\![\,1\ 0\ 0\ 2\,]\!]$ and $[\![\,1\ 0\ 0\ 3\,]\!]$. (The *object-id* of the "header area" itself is $[\![\,1\ 0\ 0\,]\!]$.) Likewise, the immediately subordinate objects of the logical object "body" are those with the *object-ids* $[\![\,3\ 2\ 0\,]\!]$, $[\![\,3\ 2\ 2\,]\!]$, $[\![\,3\ 2\ 1\,]\!]$ and $[\![\,3\ 2\ 3\,]\!]$.

Consider, for example, the ordering of the numbers for the attribute value of the "body": The "2" precedes the "1". In Fig. 16 the attribute object identifier has the value $[\![\,3\ 2\ 2\,]\!]$ for the object "1st para" and $[\![\,3\ 2\ 1\,]\!]$ for the "2nd para". Of course, the first paragraph precedes the second one and this is reflected by the value of the attribute subordinates for the "body".

(An explanation of why the *object-ids* of the two paragraphs are "rotated" might be that the author of the document created the second paragraph first and added the first one afterwards, and the ODA editor which the author used assigned *object-ids* to the objects according to the sequence in which they were created.)

The sequential logical order of the objects in the logical structure, defined by the attribute subordinates, is partially shown in Fig. 21. (The sequential layout order of the objects in the layout structure would give a similar picture.)

In Fig. 21 the upper number in each box shows the sequence number which each object has according to the sequential order, the lower number shows the value of the attribute object identifier for the object (see also Figs. 16 and 20).

As can be seen in Fig. 21, to determine the sequential order of the objects in a document, one "descends" as far as possible down the branches of the tree where the direction of descent is always guided by the first, not yet "consumed" number in sequence of the attribute subordinates. If

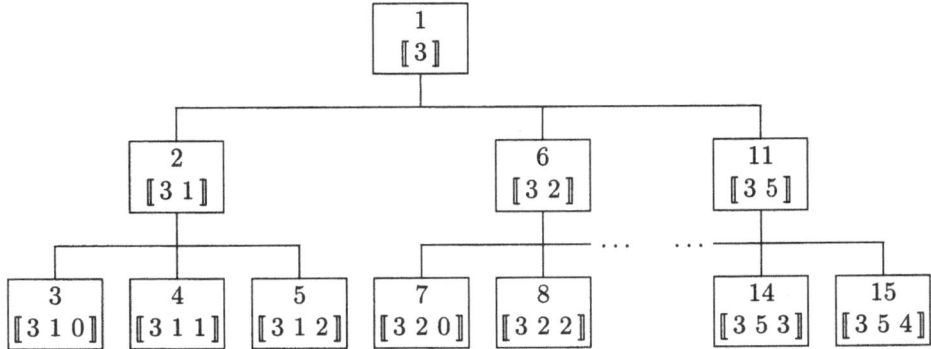

Fig. 21: Sequential logical order of the objects in the logical structure of the business letter

no number remains, the search for an "unconsumed" number continues at the immediately superior object.

The value of the attribute **generator for subordinates** is a so-called *construction expression* (see p. 82), i.e.:

generator for subordinates = '*construction expression*'

The attribute must be specified for composite object classes, i.e., for the document logical root class and document layout root class, for composite logical object classes, page set classes, composite page classes and frame classes, if the generic logical structure or generic layout structure, respectively, is a so-called *complete generator set*. If the respective structure is a so-called *factor set* the attribute is not permitted, and if the structure is a so-called *partial generator set* it may be specified or not (see Sect. 3.2.15).

The evaluation of a construction expression usually yields not a single value but a set of values (see p. 84). (The more general case of a value set discussed below includes also the case of a single value.) The significance of the attribute depends on whether the generic logical structure or generic layout structure, respectively, is a complete generator set or a partial generator set.

If the respective generic structure is a complete generator set, the value set of the attribute defines which objects may be immediately subordinate to an object that refers to the object class for which the attribute **generator for subordinates** is specified. More precisely, the objects immediately subordinate to an object, including their ordering, must be consistent with at least one value of the attribute **generator for subordinates**, specified for the object class to which the object belongs.

For clarification consider again the specific logical structure of the business letter in Fig. 3 and the generic logical structure shown in Fig. 4. For the objects and object classes the attribute values shown in Fig. 22 shall be specified. (Some of the attributes with their respective values have already been shown in Figs. 16, 17, 18, and 20; they are repeated here to provide a clearer overview on the example.)

object/ object class	object identifier/ object class identifier	subordinates/ generator for subordinates	object class
header	[3 1]	[0 1 2]	[2 0]
end	[3 5]	[1 2 3 4]	[2 1]
address	[3 1 0]	...	[2 0 1]
subject	[3 1 1]	...	[2 0 3]
date	[3 1 2]	...	[2 0 2]
greetings	[3 5 1]	...	[2 1 0]
signature	[3 5 2]	...	[2 1 2]
name	[3 5 3]	...	[2 1 1]
enclosure	[3 5 4]	...	[2 1 3]
header	[2 0]	seq [2 0 1] [2 0 2] opt [2 0 3]	—
end	[2 1]	seq [2 1 0] [2 1 2] [2 1 1] opt [2 1 3]	—
address	[2 0 1]	...	—
date	[2 0 2]	...	—
subject	[2 0 3]	...	—
greetings	[2 1 0]	...	—
signature	[2 1 2]	...	—
name	[2 1 1]	...	—
enclosure	[2 1 3]	...	—

Fig. 22: Example for explaining the attribute generator for subordinates

In Fig. 22 "..." means that the value of the attribute is of no significance for the example, and "–" means that the attribute cannot be specified for that constituent. The prefix **req** has been omitted from the values of the attribute **generator for subordinates** (see the comment on p. 84) and the prefix **sequence** has been abbreviated as **seq**, which is allowed in the ODA Standard. The constituents above the dotted line are objects, and those below are object classes.

Consider at first the object class "end" and the logical object "end" with its immediately subordinate objects. The immediately subordinate objects are those with the *object-ids* ⟦3 5 1⟧, ⟦3 5 2⟧, ⟦3 5 3⟧ and ⟦3 5 4⟧ – in exactly that order – and the object classes referenced from these objects are those with the *object-class-ids* ⟦2 1 0⟧, ⟦2 1 2⟧, ⟦2 1 1⟧ and ⟦2 1 3⟧. Consequently, ⟦⟦2 1 0⟧ ⟦2 1 2⟧ ⟦2 1 1⟧ ⟦2 1 3⟧⟧ should be a value within the set of values of the attribute **generator for subordinates** for the object class "end". As can be seen, this is in fact the case.

Consider now the object class "header" and the logical object "header" with its immediately subordinate objects. The immediately subordinate objects are those with the *object-ids* ⟦3 1 0⟧, ⟦3 1 1⟧ and ⟦3 1 2⟧ – in exactly that order – and the object classes referenced from these objects are those with the *object-class-ids* ⟦2 0 1⟧, ⟦2 0 3⟧ and ⟦2 0 2⟧. The value ⟦⟦2 0 1⟧ ⟦2 0 3⟧ ⟦2 0 2⟧⟧, however, is not within the value range of the attribute **generator for subordinates** for the object class "header", since the value range consists of the two values ⟦⟦2 0 1⟧ ⟦2 0 2⟧⟧ and ⟦⟦2 0 1⟧ ⟦2 0 2⟧ ⟦2 0 3⟧⟧. This means that the value of the attribute **subordinates** for the object "header" and the value of the attribute **generator for subordinates** for the object class "header" are inconsistent, i.e., the specific structure and the generic structure are incompatible.

If the generic logical structure or the generic layout structure is a partial generator set, the values of the attributes **subordinates** and **generator for subordinates** need not be completely compatible: there may appear subordinate objects which are not specified by the attribute **generator for subordinates**, i.e., the generic structure describes the associated specific structure only partially.

The value of the attribute **content portions** is a sequence of non-negative integers, i.e.:

content portions = ⟦ [*non-negative integer*]⁺ ⟧

This attribute can be specified for basic objects and basic object classes. It must be specified for the basic objects of the specific structures (i.e., for basic logical objects, basic pages and blocks) unless at least one of the following applies:

– The attribute **content generator** is specified. (This is only possible for basic logical objects.)
– The object refers to an object class for which the attribute **content portions** or **content generator** is specified.
– The object refers to an object class which in turn refers to an object class in a so-called *resource document* (see Sect. 3.4.3), and for the

object class in the resource document the attribute content portions is specified.

In other words, a certain piece of the actual content of a document must be associated with each basic object.

Exceptions are again the simple-structured CCITT documents (see Sect. 3.3.3). For these documents the attribute content portions can be missing though none of the three conditions above holds. The content associated with their basic objects can be identified by the structure of their data stream, and an explicit assignment with the attribute content portions is not required.

The semantics of this attribute are similar to the attribute subordinates: it specifies the constituents which are associated with a basic object or a basic object class. More precisely, if the value of the attribute object identifier or object class identifier for a particular object or object class is $[\![\, x_1 \; x_2 \, \ldots \, x_n \,]\!]$ and the value of the attribute content portions is $[\![\, y_1 \; y_2 \, \ldots \, y_m \,]\!]$, the content portions associated with the object or object class are those for which the attribute content identifier layout or content identifier logical has the value $[\![\, x_1 \; x_2 \, \ldots \, x_n \; y_1 \,]\!]$, $[\![\, x_1 \; x_2 \, \ldots \, x_n \; y_2 \,]\!]$ $\ldots$ $[\![\, x_1 \; x_2 \, \ldots \, x_n \; y_m \,]\!]$. (Which of these two attributes applies depends on whether the object or object class belongs to the logical or layout structure.) Within the sequence $[\![\, y_1 \; y_2 \, \ldots \, y_m \,]\!]$ none of the numbers can appear twice (i.e., $y_i \neq y_j$ for $i \neq j$), which should be obvious.

A content portion which is associated with an object class is called a *generic content portion*.

As with the attribute subordinates, the ordering of the numbers in the sequence is important, since it defines the sequential order of the content portions which specifies the order in which the content portions are processed by the layout process and imaging process.

3.2.7   Attributes for References

Besides the attributes for the creation of the specific and generic structures described in the previous section, there are five other attributes for establishing relations between the constituents of a document and even to constituents in a separate document. These attributes are object class, presentation style, layout style, derived from and resource.

The value of the attribute object class is an *object-class-id* (see p. 72):

object class = *object-class-id*

This attribute can be specified for all constituents of the specific structures. Furthermore, the attribute is mandatory for all constituents of the specific logical structure or for all composite constituents of the specific layout structure if the respective generic structures are complete generator sets. As mentioned already, it is used for assigning an object to an object class, for instance, to derive attribute values for an object from its object class (see Sect. 3.2.16) or to permit consistency tests between specific and generic structures.

The value of the attribute **presentation style** is either a *presentation-style-id* (see p. 72) or the value null, i.e.:

presentation style $= (presentation\text{-}style\text{-}id \mid \text{null})$

The attribute can be specified for basic objects and basic object classes, i.e., for basic logical objects, basic pages, blocks, basic logical object classes, basic page classes and block classes. It identifies a presentation style which will be applied during the layout process and imaging process to the object or to an object referring to the object class. The value null indicates that no presentation style is associated with the object or object class.

The value of the attribute **layout style** is either a *layout-style-id* (see p. 72) or the value null:

layout style $= (layout\text{-}style\text{-}id \mid \text{null})$

This attribute can be specified for all objects or object classes of the logical structure and identifies the layout style which will be applied during the layout process to the object or to an object referring to the object class. The value null indicates that no layout style is associated with the object or object class.

The value of the attribute **derived from** is either a *layout-style-id* or a *presentation-style-id* (see p. 72):

derived from $= (layout\text{-}style\text{-}id \mid presentation\text{-}style\text{-}id)$

This attribute can be specified for layout styles – in this case its value is a *layout-style-id* – or for presentation styles – in this case its value is a *presentation-style-id*. (The attribute was not included in the initially published version of the Standard; it was introduced by the Addendum on Styles.)

If the attribute is specified for a style, such a style is called a *derived style*, otherwise it is called a *root style*. A derived style "inherits" all attribute values of the style which it references by means of the attribute derived from, i.e., an attribute may not be specified directly on a particular style but indirectly on a referenced style. Furthermore, inherited attribute values may be "overwritten" by specifying a new value for the attribute. A derived style need not necessarily refer to a root style but it may also refer to another derived style, i.e., a root style and its derived styles may build a tree structure of several levels.

The attribute derived from provides a convenient method for "factorizing" the information contained in styles. For instance, the general rules describing the formatting of character content in an ODA document may be specified in a root presentation style containing all required presentation attributes for character content. Only the special rules for particular objects such as footnotes or headings need then be specified in appropriate derived presentation styles which apply to these objects.

The value of the attribute resource is a character string consisting of characters from the minimum subrepertoire of ISO 6937, Part 2, i.e.:

resource = '*ISO-6937/2 character string*'

This attribute can be specified for all object classes of the generic structures and is used for referring to object classes in a so-called *resource document* (see Sect. 3.4.3). To identify the actual object class in the resource document, the attribute resources in the document profile of the resource document must be taken into account.

3.2.8   Attributes for Describing the Content of a Document

Nine attributes are defined in Part 2 of the Standard which relate to the actual content of an ODA document. These are the attributes content architecture class, content information, alternative representation, type of coding, coding attributes, presentation attributes, bindings, content generator and logical source. Additional attributes relating to the content are described in Parts 6, 7 and 8 of the Standard where the different content architectures are defined.

Content architecture class

The attribute content architecture class is used for specifying the kind of content for the content portions associated with basic objects.

The value of the attribute content architecture class is either formatted character content architecture, processable character content architecture, formatted processable character content architecture, formatted raster graphics content architecture, formatted processable raster graphics content architecture or formatted processable geometric graphics content architecture, i.e.:

content architecture class =
    (formatted character content architecture
     | processable character content architecture
     | formatted processable character content architecture
     | formatted raster graphics content architecture
     | formatted processable raster graphics content architecture
     | formatted processable geometric graphics content architecture)

This attribute can be specified for all basic objects or object classes, i.e., for basic logical objects, basic pages, blocks, basic logical object classes, basic page classes and block classes. If it is missing for basic objects of the specific structures, a default value is derived according to the procedure described in Sect. 3.2.16. If no explicit default value can be found, the value formatted character content architecture is assumed.

The values of the attribute are not encoded as character strings but as so-called *ASN.1 object identifiers* (see p. 209). For reasons of readability, however, the meanings of these ASN.1 object identifiers are used here; the precise values are given in those parts of the Standard which specify the content architectures (see pp. 239, 273 and 305).

Content information and alternative representation

The attributes content information and alternative representation provide the simplest way to specify the actual content of a document, and content information will usually be the one which is used at most places of a document for this purpose.

The value of the attribute content information is an octet string (see p. 69) representing a piece of the content of a document. The structure of the octet string is defined in Parts 6, 7, and 8 of the Standard where the content architectures are described. Part 2 makes no assumptions concerning the make-up of the octet string, i.e.:

content information = ' *octet string* '

This attribute can be specified for content portions.

The value of the attribute **alternative representation** is a character string from a character repertoire which is specified by the attribute **alternative representation character sets** of the document profile (see p. 189):

alternative representation =
   ' *alternative representation character string* '

This attribute can be specified for content portions and defines a character string which can be imaged in lieu of the attribute **content information**. A typical application of this attribute is the following. A content portion may represent a raster image. A document containing the content portion may be sent to a receiver whose ODA system has not implemented the raster graphics content architecture, i.e., the receiving system cannot process the content portion. In this case, the attribute **alternative representation** may be used by the originator to give the receiver at least a hint as to what the content portion contains, for instance by specifying:

alternative representation = 'This is a raster image showing ...'

If the document profile of the document does not explicitly specify the attribute **alternative representation character sets**, the character repertoire used for the character string is the minimum subrepertoire of ISO 6937, Part 2.

Type of coding, coding attributes and presentation attributes

The three attributes **type of coding, coding attributes** and **presentation attributes** can be used for specifications concerning the coding and the presentation of content portions. These specifications depend, of course, on the type of content of the content portions and therefore a detailed description of the values for these attributes is given in Parts 6, 7 and 8 of the Standard. In Part 2 only rather general specifications for these attributes are given.

The value of the attribute **type of coding** is a so-called *ASN.1 object identifier* (see p. 209), depending on the content architecture of the content portion:

type of coding $= \: 'ASN.1\ object\text{-}identifier\,'$

This attribute can be specified for content portions. If it is missing a default value will be derived according to the procedure described in Sect. 3.2.16. If no explicit default value can be found one of the values specified in Parts 6, 7 or 8 as default values for this attribute will be assumed, depending on the content architecture.

(As an exception, again for compatibility with CCITT Recommendations, instead of an ASN.1 object identifier, the value of this attribute can also be the integer 0 in the case of formatted raster graphics.)

The value of the attribute **coding attributes** consists of a set of so-called *coding attributes*, depending on the content architecture.

coding attributes $=$
    $(\{\,'character\ content\ coding\ attribute\,'\}$
    $|\,\{\,'raster\ graphics\ content\ coding\ attribute\,'\}$
    $|\,\{\,'geometric\ graphics\ content\ coding\ attribute\,'\}\:)$

This attribute can be specified for content portions. The permitted coding attributes, their classification as non-mandatory or defaultable and their possible default values are specified in Parts 6, 7 and 8 of the Standard (see pp. 240 and 305 and Sect. 7.2.1).

The value of the attribute **presentation attributes** consists of one or more sets of so-called *presentation attributes* for the different content architecture classes:

presentation attributes $=$
    $\{[\{\,'character\ content\ presentation\ attribute\,'\}]$
    $[\{\,'raster\ graphics\ content\ presentation\ attribute\,'\}]$
    $[\{\,'geometric\ graphics\ content\ presentation\ attribute\,'\}]\}$

This attribute can be specified for basic page classes, block classes, basic pages, blocks and presentation styles. If the attribute is missing for a basic page or a block, or if a particular presentation attribute which is needed during the layout process or imaging process is not contained in the value of the attribute, a default value is determined according to the procedure described in Sect. 3.2.16. If no explicit default value can be found the default value defined in Parts 6, 7 and 8 of the Standard is assumed (see Sects. 6.2.1, 7.2.2 and 8.2.2).

The specification of more than one set of presentation attributes is not sensible for basic page classes, block classes, basic pages and blocks since their associated content portions can be of only one content architecture class, specified by means of the attribute content architecture class; only the presentation attributes relating to this content architecture class can be applied. For presentation styles, however, it can be necessary to specify several sets of presentation attributes, each one relating to a different content architecture class, since a presentation style may be referenced from several basic objects with associated content portions of different content architecture classes.

Bindings, content generator and logical source

The three attributes bindings, content generator and logical source are closely related to each other and permit the creation of pieces of content of a document during the layout process. Such a creation of content is needed, for instance, for numbering the pages of a document or for adding headlines containing the title of the current section automatically during the layout process.

The value of the attribute bindings is a set of pairs where each pair consists of a *binding name* and a *binding value*. A *binding name* is a character string from the minimum subrepertoire of ISO 6937, Part 2. A *binding value* is an object identifier expression, string expression or numeric expression if the attribute is specified for object classes of the generic logical or layout structure or for logical objects, and it is a string literal, an integer or an *object-id* if the attribute is specified for layout objects (see Sect. 3.2.3). This can be summarized as:

bindings = {[' *binding name* ' ' *binding value* ']$^+$}

binding name := ' *ISO-6937/2 character string* '

binding value :=
   (' *object identifier expression* ' | ' *string expression* '
   | ' *numeric expression* ' | ' *string literal* ' | *integer* | *object-id*)

This attribute can be specified for all kinds of objects or object classes. If the attribute is missing or if a particular binding name is missing in the value of the attribute, a default value will be derived according to procedure described in Sect. 3.2.16. If no explicit default value can be found within the document, the binding value for the binding name is undefined.

With the attribute **bindings** the concept of variables as known from programming languages is introduced, though in a rather restricted way. The binding name corresponds to the name of a variable, and the binding value corresponds to the value of a variable or, in the case of expressions, to the specification of a rule to say how this value is "computed".

The "declaration" of such a variable and the assignment of a value, or the definition of a rule for "computing" the value, is made by specifying the attribute **bindings** with a pair "'*binding name*' '*binding value*'". As mentioned, such a declaration can be made for any object or object class. A reference to a binding value for a particular binding name, i.e., an access to the value of such a variable, can only be made with the attribute **content generator**.

The value of the attribute **content generator** is a string expression:

content generator = '*string expression*'

This attribute can be specified for basic object classes and basic logical objects, i.e., for basic page classes, block classes, basic logical object classes and basic logical objects. The string expression is evaluated during the layout process and the resulting character string is a piece of content associated with the basic object for which the attribute is specified. Usually, the character string will represent character content; however, an interpretation as raster graphics or geometric graphics content is also permitted in the Standard. The interpretation of the character string depends on the content architecture class associated with the basic object (see p. 99).

The attribute **content generator** is ignored if the object has more than one associated content portion, or in case exactly one is associated with it, if for this content portion the attribute **content information** is specified (see p. 99).

Simplifying the situation slightly, there are essentially two possibilities for assigning content to a basic object: firstly, by specifying the attribute **content information** for a content portion which is associated with the object, and secondly, by specifying the attribute **content generator** for the object. In both cases this could also be done in the generic structures if the object belongs to an object class. The precise rules for determining the content associated with a basic object are given in Sect. 3.6.2, p. 164ff., and Sect. 3.6.3, p. 167ff.

As an example for the application of the attributes **bindings** and **content generator**, let us consider the automatic creation of chapter numbers. In a particular document, the logical objects immediately subordinate to the document logical root shall be "chapters" which in turn shall have the subordinate objects "chapter numbers" (a basic logical object) and "chapter text". The substructure of "chapter text" is of no relevance for the example. Figure 23 shows an example of the logical structure of such a document.

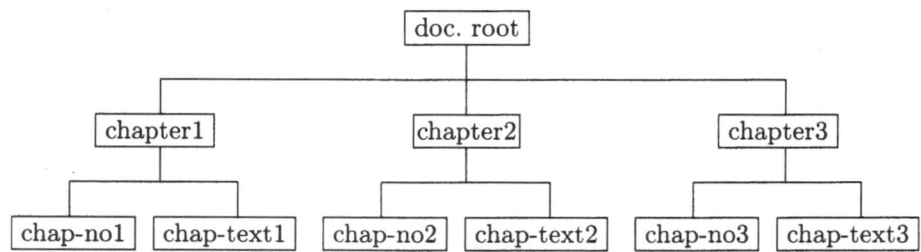

Fig. 23: Specific logical structure of a document

The automatic creation of the chapter numbers during the layout process could be done as follows: A "variable" is needed which is initialized with the value 0 and incremented by 1 each time a new chapter starts. The "declaration" of this variable and its initialization should be done at the document logical root level by means of the attribute **bindings**, i.e., the attribute **bindings** is specified for the document logical root with the following value:

bindings = {(chapno 0)}

As a consequence, the binding value is in this case the integer 0. For the basic logical objects representing the chapter numbers, i.e., for "chapno1", "chap-no2" and "chap-no3" the attribute **bindings** would be specified as:

bindings =
{(chapno (INCREMENT(PRECEDING(CURRENT-OBJECT) chapno)))}

(Parentheses have been inserted into the expression to make the structure of the expression explicit.) In this case, the binding value is a numeric expression: the function INCREMENT, applied to the numeric expression (PRECEDING(CURRENT-OBJECT) chapno) (see p. 75). This is, in turn, a binding reference with the binding name **chapno** and the binding reference expression PRECEDING(CURRENT-OBJECT) referring to the object which precedes the one for which the attribute bindings is specified, in the sequential logical order (see p. 80).

The access to the "variable" chapno is performed with the attribute content generator which would have the value

content generator = MAKE-STRING(CURRENT-OBJECT chapno)

for the logical objects "chap-no1", "chap-no2" and "chap-no3". This value is a string expression: the string function MAKE-STRING, applied to the numeric expression CURRENT-OBJECT chapno (see p. 74). This expression, in turn, is a binding reference, referring to the binding value for the binding name chapno of the object for which the attribute content generator is specified. Somewhat simplified, this value of the attribute content generator takes the numeric value of the variable chapno, as specified for the respective object, and transforms it into a character string.

During the layout process, for instance, when the logical object "chap-no2" is to be processed, the following will happen: It is observed, that no content portion is associated with the basic logical object "chap-no2", but the attribute content generator is specified for this object. Therefore, the content which belongs to this basic logical object is the determined by the evaluation of the value of this attribute. The evaluation requires the determination of the binding value for the binding name chapno which is found within the value of the attribute bindings for "chap-no2".

The binding value for this binding name essentially specifies that it is the value of the binding name chapno at the preceding object, incremented by 1. At the preceding object "chapter2", however, there is no such binding name specified and therefore the search for the binding name is continued in the direction opposite to the sequential logical order (see p. 80), i.e., at the object "chap-text1". (This would most likely be a composite logical object, i.e., the search would first be performed at the subordinate objects of "chap-text1".) Again, the binding name is not found and the search continues at the object "chap-no1". Here the binding name is present within the value of the attribute bindings, but again it is specified that it is the value of the binding name chapno at the preceding object, incremented by 1. At the preceding object "chapter1", the binding name chapno is not found, but it is finally identified at the document logical root with the value 0. This value is then incremented twice by 1, i.e., the value for the binding name chapno is 2 for the object "chap-no2" and the content associated with this basic logical object is therefore the character string "2".

If a generic logical structure were present, which would be rather natural for this example, the attributes bindings and content generator could be factorized into an object class, e.g., "chap-no", so the attributes need not be repeated for each individual object representing a chapter number.

The attribute logical source is called a *formatting attribute* in the ODA Standard which might be slightly misleading. Besides content information and content generator it is the third attribute by which pieces of content of a document can be created.

The value of the attribute logical source is a *logical object class identifier*:

logical source = *logical-object-class-id*

This attribute can be specified for frame classes and has the following effect: Whenever a frame belonging to the frame class is created during the layout process, also a logical object belonging to the object class specified by the value of the attribute is created. If for this object class the attribute generator for subordinates is specified, its value will be determined and the corresponding subordinate logical objects created. These two steps – evaluation of the attribute generator for subordinates and creation of the corresponding subordinate logical objects – may be repeated several times until finally only basic logical objects are left.

For a better understanding of this process consider the example in Fig. 24 where for five object classes their *object-class-ids* and values for the attribute generator for subordinates are specified. It is also indicated whether the object classes are composite or basic.

object-class-id	generator for subordinates	type of object class
⟦2 4⟧	seq ⟦2 4 0⟧ ⟦2 4 1⟧	composite
⟦2 4 0⟧	(not specified)	basic
⟦2 4 1⟧	seq ⟦2 4 1 0⟧ ⟦2 4 1 1⟧	composite
⟦2 4 1 0⟧	(not specified)	basic
⟦2 4 1 1⟧	(not specified)	basic

Fig. 24: Excerpt from the generic logical structure of a document

If a frame class which refers to the object class with the *object-class-id* ⟦2 4⟧ by means of the attribute logical source is referenced during the layout from a particular frame, two composite and three basic logical objects which are connected in a tree structure are created, according to the values of the attribute generator for subordinates of the logical object

object-id	subordinates	object class
⟦3 7 6⟧	⟦0 1⟧	⟦2 4⟧
⟦3 7 6 0⟧	(not specified; basic logical object)	⟦2 4 0⟧
⟦3 7 6 1⟧	⟦0 1⟧	⟦2 4 1⟧
⟦3 7 6 1 0⟧	(not specified; basic logical object)	⟦2 4 1 0⟧
⟦3 7 6 1 1⟧	(not specified; basic logical object)	⟦2 4 1 1⟧

Fig. 25: Logical objects created by means of the attribute **logical source**

classes. These generated logical objects might have the *object-ids* and values for the attributes **subordinates** and **object class** as shown in Fig. 25.

The root node of this subtree has therefore the *object-id* ⟦3 7 6⟧ and belongs to the object class with the *object-class-id* ⟦2 4⟧. This object has two subordinate objects with the *object-ids* ⟦3 7 6 0⟧ (a basic logical object) and ⟦3 7 6 1⟧ (a composite logical object) which belong to the object classes with the *object-class-ids* ⟦2 4 0⟧ and ⟦2 4 1⟧, respectively. The composite logical object ⟦3 7 6 1⟧ has to subordinate basic logical objects with the *object-ids* ⟦3 7 6 1 0⟧ and ⟦3 7 6 1 1⟧ which belong to the object classes ⟦2 4 1 0⟧ and ⟦2 4 1 1⟧.

The following should be noted: The evaluation of the attribute **generator for subordinates** must yield exactly one value since otherwise it would not be uniquely specified which logical objects should be created. A set of values for the attribute **generator for subordinates**, for instance, the ones shown on p. 84, is not permitted in the case that an evaluation of the attribute **generator for subordinates** is caused by the attribute **logical source**.

The basic logical objects created by this process now have some sort of associated content. For the basic logical object classes to which the objects belong, the attributes **content portions** or **content generator** are specified, referring to generic content portions or generating content directly. This content will be laid out in the frame which caused the creation of the logical objects.

The logical objects created by this process are called *temporary logical objects*. They are not considered genuine objects of the specific logical structure, and in particular they are not part of the interchanged data stream. When talking about logical objects, these temporary objects are never meant except when they are explicitly mentioned. However, the layout objects created by this process (subordinate frames and blocks with associated content portions) become genuine objects of the specific layout structure and appear also in the interchanged data stream.

To summarize in a slightly generalized manner, by means of the attribute logical source temporary logical objects of the specific logical structure can be created which were not present in advance. The basic logical objects generated in this way have associated content, as usual, which will then be laid out by the layout process, i.e., the layout process can create content of a document.

A typical application of this attribute is, for instance, the automatic creation of headlines or footlines during the layout process. This could be done as follows: For each page class in the generic layout structure two subordinate frame classes are specified: one for the actual content on the page and one for the headline. For the frame class representing the headline, the attribute logical source is specified which refers to an object class in the generic logical structure. For this logical object class, which shall be basic, the attribute content generator is specified whose evaluation yields the text which shall be shown in the headline.

Whenever the layout process creates a new page, a frame belonging to the frame class for the headline is automatically created and also a content portion which contains the laid out text created by the attribute content generator. (Of course, the logical object class referenced by the attribute logical source may also be composite if the content in the headline is composed of several pieces, for instance, if it contains a company logo, a chapter number and a page number. The actual content of the headline is then associated with basic logical objects directly or indirectly subordinate to the composite logical object.)

3.2.9   Attributes for the Inclusion of Comments

There are three attributes – application comments, user-readable comments and user-visible name – which can be used to add comments to constituents. Opposite to all other attributes, the semantics of these three attributes is not defined precisely in the ODA Standard. In particular, it is not specified how these attributes shall be processed by an ODA implementation and their usage is left to the authors of ODA documents. They will usually be used to add information to constituents which may be read by humans but not processed by an ODA system.

The values of the attributes user-readable comments and user-visible name consist of a character string from a character repertoire which is specified by the attribute comments character sets in the document profile (see p. 189). If the attribute is missing in the document profile the minimum subrepertoire of ISO 6937, Part 2, extended by the control functions *"space"*, *"carriage return"* and *"line feed"* is used:

user-readable comments = *comments character string*

user-visible name = *comments character string*

The attributes user-readable comments and user-visible name can be specified for all constituents of the specific and generic structures, for layout styles and presentation styles. If they are missing for objects of the specific structures a default value will be determined according to the procedure described in Sect. 3.2.16. If no explicit default value can be found within the document an empty character string is assumed.

The value of the attribute application comments is any sequence of bytes (see p. 69):

application comments = *octet string*

This attribute can also be specified for all constituents of the specific and generic structures, but not for layout styles and presentation styles. If it is missing for objects of the specific structures a default value will be determined according to the procedure described in Sect. 3.2.16. If no explicit default value can be found within the document an empty octet string is assumed.

The names of the attributes give some hints for their recommended usage. The attribute user-visible name shall be used for adding character strings as "names" to the constituents. The actual "names" of the constituents in an ODA document are sequences of numbers (see pp. 69 – 73) which do not reflect the semantic role of the constituents in a particular document.

By using the attribute user-visible name and specifying, for instance, "user-visible name=`'address'`" for an object class which shall represent the address of a letter, a human reader can recognize the semantic role of the object class in the generic structure of a document by looking at the value of the attribute user-visible name. (This requires, of course, that the author uses an ODA system which provides a convenient way for doing this.) Though the term "address" is not defined in the ODA Standard, a human reader will certainly have an intuitive knowledge of what an address is.

Further comments can be given with the attributes application comments and user-readable comments. The first one should be used for application dependent comments, the second one for application independent comments concerning the respective constituents, but the specifications of the Standard are rather vague. This is acceptable since none of the three attributes is of importance for the layout or imaging process.

3.2.10   Attributes for Controlling the Layout Process

The ODA Standard defines several attributes which control the layout process. These attributes are, in alphabetical order, balance, block alignment, border, concatenation, dimensions, fill order, indivisibility, layout category, layout object class, layout path, new layout object, offset, permitted categories, position, same layout object, separation and synchronization.

These attributes are discussed in this section, classified into several groups according to their functionality. Such a classification is not always unique, however, since the same layout feature is sometimes achievable by different attributes.

Layout path and fill order

The attributes layout path and fill order specify the directions in which the layout objects (frames and blocks) shall be arranged on the presentation medium during the layout process.

The value of the attribute layout path is either 0°, 90°, 180° or 270°, i.e.:

layout path = (0° | 90° | 180° | 270°)

This attribute can be specified for frames and frame classes. If it is missing for a frame a default value will be derived according to the procedure described in Sect. 3.2.16. If no explicit default value can be found 270° is assumed.

The value of this attribute defines the direction of progression – measured counterclockwise from the left-to-right direction – into which are allocated the immediately subordinate objects of the object for which the attribute is specified, during the layout process. This direction is called the *layout path*. However, this attribute is only applied if the position of the subordinate objects is not yet fixed, i.e., the value of the attribute position for the immediately subordinate objects must be variable position (see p. 117). For clarification consider Fig. 26.

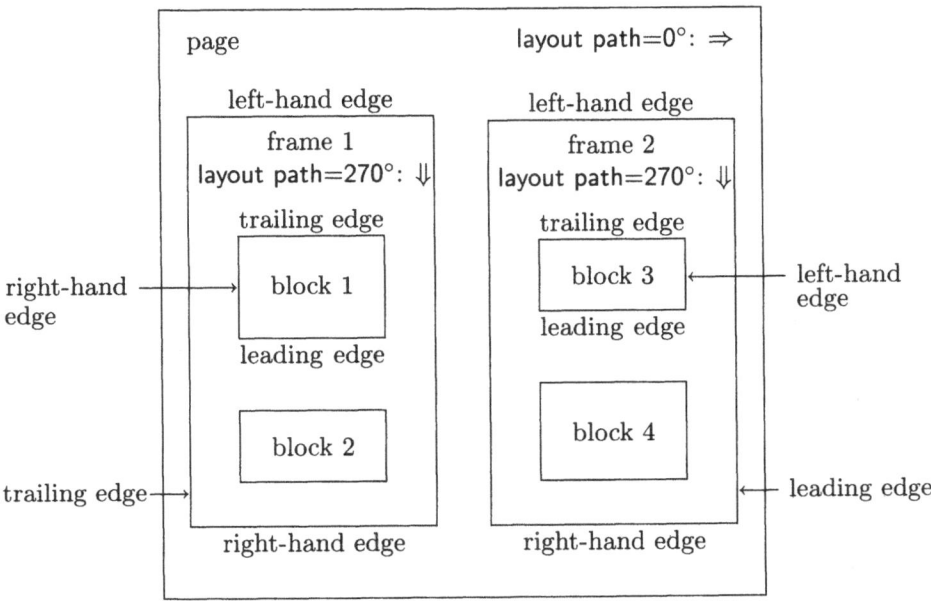

Fig. 26: Example for the layout path and the names of the edges

Seven layout objects are shown in this example: a page with two sub-ordinate frames which each in turn contain two blocks. For the page the value of the attribute layout path is 0°, i.e., the second frame is placed at the right side of the first one. For the two frames the value of the attribute layout path is 270°, i.e., the blocks within the frames are allocated below each other. These specifications for the layout paths reflect a two-column layout format.

It should be noted that the layout path can also be 90° ("from bottom to top") and 180° ("from right to left"). Therefore, the ODA Standard is also applicable to documents where the writing direction is not from left to right and top to bottom as in English documents.

Figure 26 shows several names for the edges of layout objects. Each layout object has four edges called *leading edge, trailing edge, left-hand edge* and *right-hand edge*. It should be noted that the names of the edges depend on the direction of the layout path: leading edge and trailing edge are always orthogonal, left-hand edge and right-hand edge parallel to the layout path.

The value of the attribute fill order is either normal order or reverse order:

fill order = (normal order | reverse order)

This attribute can be specified for layout styles and is used during the layout process for the processing of basic logical objects. Its value is derived according to the procedure described in Sect. 3.2.16; its default value is normal order. It associates a so-called *fill order* with each basic logical object.

The attribute defines where the blocks which contain the content associated with basic logical objects shall be allocated within the immediately superior layout object (a frame or a basic page), relative to the layout path of the superior layout object.

All blocks for which the value normal order applies are laid out in the direction of the layout path corresponding to the sequential order of the logical object whose content they contain. The trailing edge of the first of these blocks neighbors the edge at the upper end of the superior layout object.

All blocks for which the value reverse order applies are also laid out in the direction of the layout path corresponding to the sequential order of the logical object whose content they contain. However, the leading edge of the first of these blocks neighbors the edge at the lower end of the superior layout object. Figure 27 shows an example for this attribute.

This example shows a frame with four subordinate blocks. The sequential order may agree with the numbering of the blocks and the layout path is from top to bottom. For the first and fourth block the fill order is normal order, for the second and third block reverse order. (It should be noted that the attribute fill order is not specified directly on the blocks but its value is determined indirectly by the layout styles which are referenced from the basic logical objects whose content is laid out in the blocks.) The first and fourth block are then placed at the "upper end", the second and third at the "lower end" of the superior frame.

A typical application of this attribute is the processing of footnotes within a document. For instance, the example in Fig. 27 could be interpreted as showing in blocks 1 and 4 the ordinary text of a text column on a page, and in blocks 2 and 3 two footnotes which are referenced from text within the column.

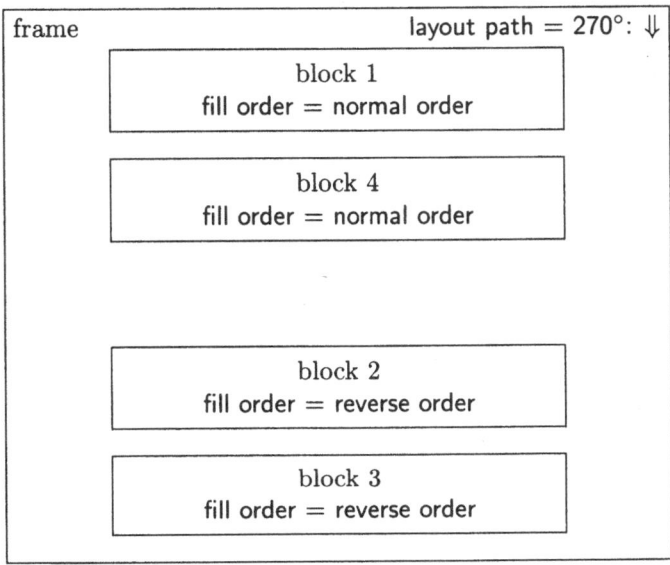

Fig. 27: Example for the attribute fill order

Block alignment, offset and separation

The three attributes block alignment, offset and separation supply additional information on how layout objects (frames and blocks) shall be allocated relative to each other.

The value of the attribute block alignment is either right-hand aligned, left-hand aligned, centred or null:

block alignment = (right-hand aligned | left-hand aligned | centred | null)

This attribute can be specified for layout styles and is used during the layout process for the processing of basic logical objects. Its value is derived according to the procedure described in Sect. 3.2.16; its default value is right-hand aligned.

The attribute specifies the alignment of blocks which contain the content associated with basic logical objects within the available area (see p. 170) of their immediately superior layout object. The values left-hand aligned, right-hand aligned and centred relate to the direction orthogonal to the layout path of the superior layout object. For the allocation of the blocks the values of the attributes offset (see p. 114) and border (see p. 121) are also taken into account.

Figure 28 shows examples for the three values of the attribute, assuming that for the blocks the parameters of the attribute offset have the value 0 and that neither for the blocks nor the frames is the attribute border specified.

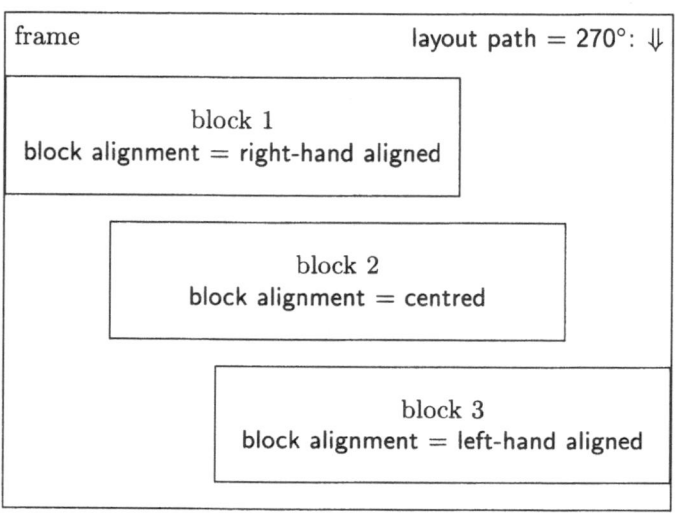

Fig. 28: Example for the effect of the attribute block alignment

It should be noted that the values right-hand aligned and left-hand aligned do not reflect the usual meaning of "right aligned" and "left aligned" layout. The directions "left" and "right" in the attribute values are reversed in the ODA Standard.

The value of the attribute offset is substructured into four parameters with the names leading offset, trailing offset, left-hand offset and right-hand offset where the value of each parameter is a non-negative integer.

offset =
    {[leading offset = *non-negative integer*]
    [trailing offset = *non-negative integer*]
    [right-hand offset = *non-negative integer*]
    [left-hand offset = *non-negative integer*]}

This attribute can be specified for layout styles and is used during the layout process for the processing of basic logical objects. Its value is derived according to the procedure described in Sect. 3.2.16; as its default value each parameter has the value 0.

The values of the parameters specify the minimum distance of the four edges of the block which contains the content of the basic logical object for which the attribute applies, to the respective edges of the superior layout object (frame). The distances are specified in scaled measurement units. For instance, the parameter left-hand offset specifies the distance of the left-hand edge of the block to the edge of the frame. Figure 29 shows the effect of the attribute graphically.

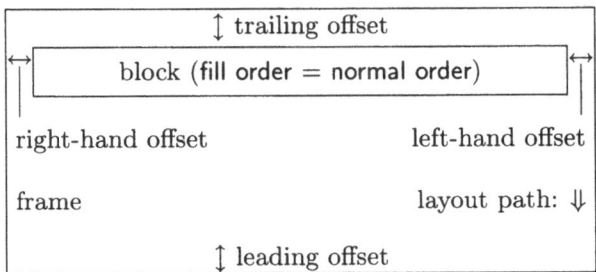

Fig. 29: Effect of the attribute offset

In Fig. 29 the four parameter trailing offset, leading offset, left-hand offset and right-hand offset are shown. It should be noted that the position of the four distances depends on the direction of the layout path. In this example, the actual distances of the trailing edge, left-hand edge and right-hand edge are equal to the minimum distances specified by the respective parameters of the attribute offset. For the leading edge of the block, however, the actual distance is greater than the minimum distance specified by the parameter leading offset, since the vertical size of the frame is greater than the minimum size required by the block, including the trailing and leading offsets. The usage of this attribute to specify the layout of a document should be obvious.

The value of the attribute separation is substructured into the three parameters leading edge, trailing edge and centre separation where the value of each parameter is a non-negative integer:

separation =
  {[leading edge = *non-negative integer*]
  [trailing edge = *non-negative integer*]
  [centre separation = *non-negative integer*]}

This attribute can be specified for layout styles and is used during the layout process for the processing of basic logical objects. Its value is derived according to the procedure described in Sect. 3.2.16; as its default value each parameter has the value 0.

The value of the parameter **leading edge** specifies the minimum distance of the leading edge of the block which contains the content of the basic logical object for which the attribute applies, to the trailing edge of the block which immediately follows in the specific layout structure and has the same fill order (see p. 112). The distance is specified in scaled measurement units, which also holds for the two other parameters.

The value of the parameter **trailing edge** specifies the minimum distance of the trailing edge of the block which contains the content of the basic logical object for which the attribute applies, to the leading edge of the block which immediately precedes in the specific layout structure and has the same fill order.

The value of the parameter **centre separation** specifies the minimum distance of blocks with different fill orders. Figure 30 shows an example for the usage of this attribute.

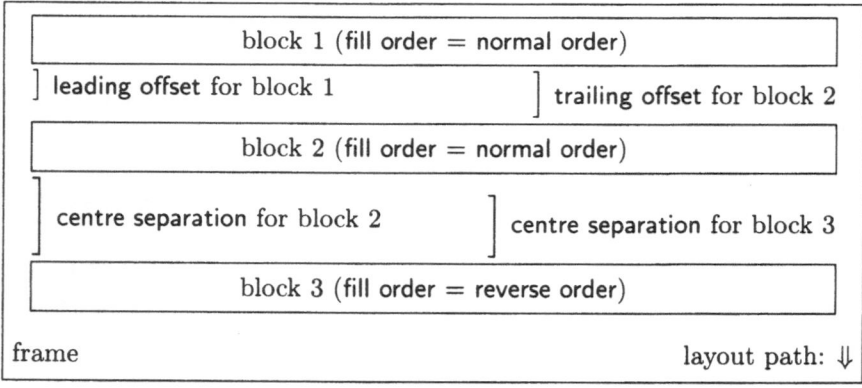

Fig. 30: Example for the attribute **separation**

For the first two blocks at the top the attribute **fill order** has the value **normal order**, for the block at the bottom the value **reverse order**. Several values of the parameters **leading offset**, **trailing offset** and **centre separation** for the three layout objects are shown graphically in Fig. 30. The distance between the first and second block is the value of the parameter **trailing offset** for the second block since this value is greater than the value of the parameter **leading offset** for the first block. The distance between the second and third block is (at least) the value of the parameter **centre separation** for the second block since this value is greater than the value of the parameter **centre separation** for the third block.

Position, dimensions and border

The two attributes position and dimensions specify, as their names indicate already, the position and size of layout objects; the attribute border specifies whether – and if so how – a border shall be drawn around a layout object. The value of all three attributes may have a rather complex structure.

The value of the attribute position is either the parameter fixed position or variable position. For the parameter fixed position the sub-parameters horizontal position and vertical position can be specified whose values are non-negative integers. For the parameter variable position the sub-parameters offset, separation, alignment and fill order are possible.

The values of the sub-parameter offset can be the sub-sub-parameters leading offset, trailing offset, left-hand offset and right-hand offset whose values are non-negative integers. For the sub-parameter separation the sub-sub-parameters leading edge, trailing edge and centre separation can be specified whose values are also non-negative integers. The value of the sub-parameter alignment is either right-hand aligned, centred or left-hand aligned, and the value of the sub-parameter fill order is either normal order or reverse order. We therefore have:

position =
   (fixed position = {[horizontal position = *non-negative integer*],
                   [vertical position = *non-negative integer*]}
   | variable position =
      {[offset = {[leading offset = *non-negative integer*],
             [trailing offset = *non-negative integer*],
             [left-hand offset = *non-negative integer*],
             [right-hand offset = *non-negative integer*]}],
      [separation = {[leading edge = *non-negative integer*],
               [trailing edge = *non-negative integer*],
               [centre separation = *non-negative integer*]}],
      [alignment = (right-hand aligned | centred | left-hand aligned)],
      [fill order = (normal order | reverse order)]})

This attribute can be specified for frames and blocks, frame classes and block classes. The parameter variable position, however, may only be used for a frame class which is not immediately subordinate to a page class. (More precisely, there exists no page class in the document which refers to such a frame class by means of the attribute generator for subordinates.) Otherwise, the value of the attribute is always fixed position.

If the attribute is missing for frames or blocks its value is derived according to the procedure described in Sect. 3.2.16. If no explicit value can be found, the parameter fixed position with the sub-parameter values horizontal position = 0 and vertical position = 0 is assumed. All sub-parameters and sub-sub-parameters may be missing. The default values defined in the Standard are right-hand aligned for the sub-parameter alignment, normal order for the sub-parameter fill order, and 0 in all other cases.

The attribute defines the position of a block or a frame relative to its immediately superior layout object (page or frame). For the parameter fixed position, the sub-parameters horizontal position and vertical position specify the horizontal and vertical distance of the layout object to the edges of its superior layout object in scaled measurement units (see p. 148). Figure 31 shows the meaning of the parameter fixed position graphically.

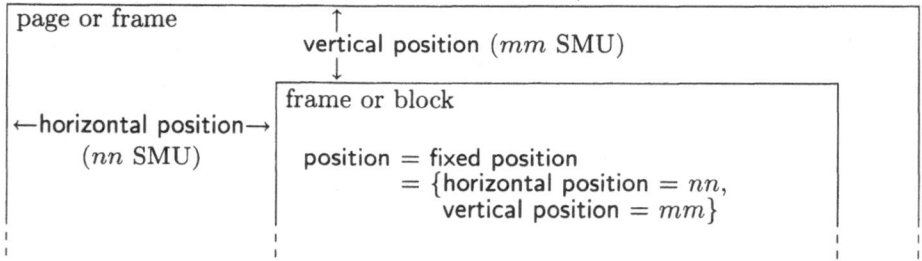

Fig. 31: Visualization of the parameter fixed position for the attribute position

If the parameter variable position is specified for the attribute position – as mentioned above, this is only possible for certain types of frame classes – the sub-parameters offset, separation, alignment and fill order specify a set of rules according to which the layout process shall determine the position of frames belonging to the frame class.

The four sub-sub-parameters leading offset, trailing offset, left-hand offset and right-hand offset of the sub-parameter offset specify the minimum distance from the four edges of a frame belonging to the frame class to the respective edges of the superior frame. The distances are specified in scaled measurement units. The effect of this sub-parameter is essentially identical to the attribute offset which can be specified for layout styles (see p. 114).

The sub-sub-parameters leading edge, trailing edge and centre separation of the sub-parameter separation specify minimum distances between a frame belonging to the frame class and its two adjacent frames, i.e., the one that precedes and the one that follows immediately in the sequential order of the layout objects. The distances are again specified in scaled measurement units. The effect of this sub-parameter is essentially identical to the attribute separation which can be specified for layout styles (see p. 115).

The sub-parameter alignment specifies the alignment of a frame belonging to the frame class within the area of its immediately superior frame. The values left-hand aligned, right-hand aligned and centred relate to the direction orthogonal to the layout path of the superior frame. For the allocation of the frame the values of the attributes offset (see p. 114) and border (see p. 121) are also taken into account. The effect of this sub-parameter is essentially identical to the attribute block alignment which can be specified for layout styles (see p. 113).

The sub-parameter fill order specifies how a frame belonging to the frame class is to be positioned in its immediately superior frame relative to the direction of the layout path of that frame. The effect of this sub-parameter is essentially identical to the attribute fill order which can be specified for layout styles (see p. 112).

The attribute dimensions has the two parameters horizontal dimension and vertical dimension whose value is one of the four sub-parameters fixed dimension, maximum size, rule A or rule B. As an alternative for the parameter vertical dimension, the sub-parameter variable page height may be specified. The sub-parameter maximum size has the value applies, and the values of fixed dimension and variable page height are positive integers. The sub-parameters rule A and rule B both have the (optional) sub-sub-parameters minimum dimension and maximum dimension whose values are also positive integers. We therefore have:

```
dimensions =
   {[horizontal dimension =
           (fixed dimension = positive integer | maximum size = applies
           | rule A = {[minimum dimension = positive integer]
                       [maximum dimension = positive integer]}
           | rule B = {[minimum dimension = positive integer]
                       [maximum dimension = positive integer]})]
    [vertical dimension =
           (fixed dimension = positive integer | maximum size = applies
           | variable page height = integer
           | rule A = {[minimum dimension = positive integer]
                       [maximum dimension = positive integer]}
           | rule B = {[minimum dimension = positive integer]
                       [maximum dimension = positive integer]})]}
```

This attribute can be specified for pages, frames, blocks, page classes, frame classes and block classes. When specified for objects in the specific layout structure it specifies their horizontal and vertical size; when specified for object classes in the generic layout structure it specifies rules according to which the size of objects belonging to the object class shall be determined. All numbers denote scaled measurement units.

If the attribute is missing for frames or blocks or if values for the parameters are missing, the attribute value or the values for the parameters are derived according to the procedure described in Sect. 3.2.16. If no explicit value can be found, the following values are assumed: the parameters horizontal dimension and vertical dimension have the value fixed dimension where the sizes of these dimensions are for frames and blocks the sizes of their superior layout object (frame or page) and for pages the size of an ISO A4 page (see p. 177).

With the sub-parameter fixed dimension the height and width of the layout object is specified explicitly. The sub-parameter maximum size with the value applies specifies that the dimension concerned shall take its default value.

For pages, frames and blocks, only the sub-parameter fixed dimension is permitted since as one of the results of the layout process all dimensions of the layout objects are determined. (An exception are basic pages for which the sub-parameter variable page height may be specified; this exception is permitted for compatibility with existing CCITT Recommendations. This sub-parameter cannot be used for other layout objects.)

The sub-parameters rule A and rule B are only allowed for frame classes which are not immediately subordinate to a page class. (More precisely: there exists no page class which refers to the frame class by means of the attribute generator for subordinates.) These two sub-parameters denote rules according to which the dimensions of frames belonging to the frame class are to be determined.

The specification of rule B means that the dimension concerned shall take the minimum value which is necessary to contain all the subordinate layout objects. In other words, if the layout process has been carried out for all the immediately subordinate layout objects of such a frame and their horizontal or vertical sizes are known, these dimensions are added up, thus yielding the dimension of the frame.

The specification of rule A gives a similar result. However, in this case not all immediate subordinate are taken into account but only the dimension of that immediate subordinate layout object which is the first one in the sequential logical order (see p. 91) and contains a piece of content of the document. The specification of rule A is only permitted in the direction of the layout path.

For the two sub-parameters rule A and rule B the (optional) sub-sub-parameters minimum dimension and maximum dimension can be specified. The value of minimum dimension specifies that at least this size should be taken for the dimension concerned, even if the subordinate layout objects would require a smaller value. The value of maximum dimension specifies that at most this size should be taken for the dimension concerned, even if this value is not sufficient for containing the subordinate layout objects.

In all these cases, it is, of course, necessary that a layout objects fits into its superior layout object. A block, for instance, can never be wider or higher than its superior frame.

The value of the attribute border consists of the four (optional) parameters left-hand edge, right-hand edge, trailing edge and leading edge. Each parameter has either the value null or one or more of the sub-parameters border line width, border line type and border freespace width. The values of the sub-parameters border line width and border freespace width are non-negative integers. The value of border line type is either solid, dashed, dot, dash-dot, dash-dot-dot or invisible.

```
border =
   {[left-hand edge =
         (null | {[border line width = 'non-negative integer']
                  [border freespace width = 'non-negative integer']
                  [border line type = (solid | dashed | dot | dash-dot
                                       | dash-dot-dot | invisible) ]})]
    [right-hand edge =
         (null | {[border line width = 'non-negative integer']
                  [border freespace width = 'non-negative integer']
                  [border line type = (solid | dashed | dot | dash-dot
                                       | dash-dot-dot | invisible) ]})]
    [trailing edge =
         (null | {[border line width = 'non-negative integer']
                  [border freespace width = 'non-negative integer']
                  [border line type = (solid | dashed | dot | dash-dot
                                       | dash-dot-dot | invisible) ]})]
    [leading edge =
         (null | {[border line width = 'non-negative integer']
                  [border freespace width = 'non-negative integer']
                  [border line type = (solid | dashed | dot | dash-dot
                                       | dash-dot-dot | invisible) ]})]}
```

This attribute can be specified for frames, blocks, frame classes, block classes, and presentation styles and determines whether or not a border is to be drawn around the layout object for which it is specified. If the attribute is missing for frames or blocks, a default value is derived according to the procedure described in Sect. 3.2.16. If no explicit value can be found, for all parameters the sub-parameters border line width and border freespace width have the value 0 and the sub-parameter border line type has the value solid, i.e., no (visible) border is drawn around the layout object.

The four parameters define the look of the border on the left-hand edge, right-hand edge, trailing edge and leading edge (see p. 111). The value of the sub-parameter border line width specifies the width of the border line in scaled measurement units (see p. 148). The value of the attribute border freespace width defines the distance between the border line and the edge of the layout object and the sub-parameter border line type specifies the fashion in which the border line shall be drawn.

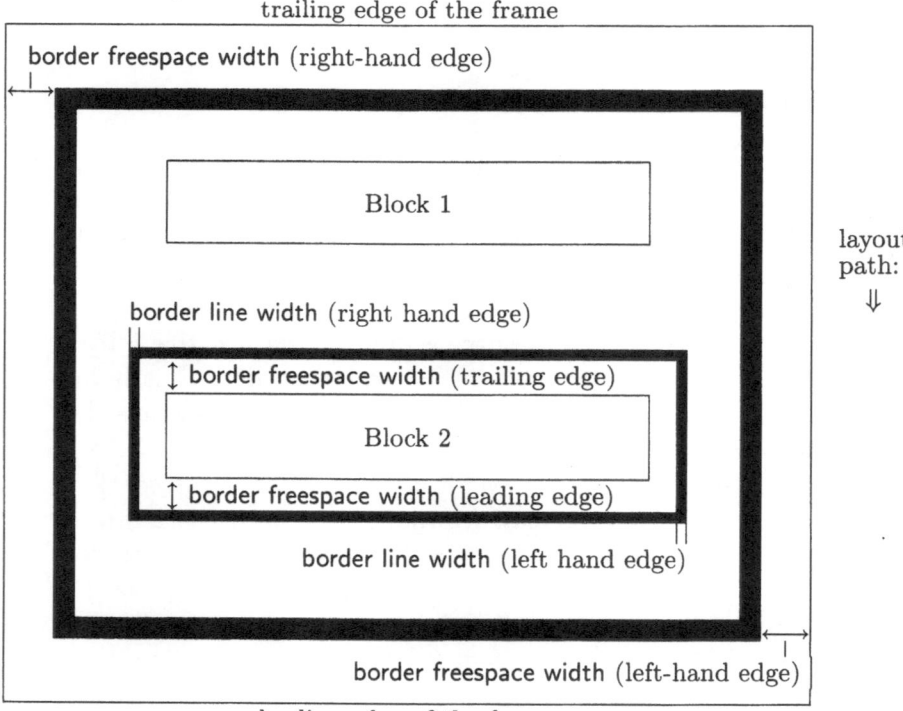

Fig. 32: Borders around blocks and frames

It should be noted that for frames the border is drawn within the area of the frame whereas for blocks the border is drawn outside the area of the block. As a consequence, the available area for displaying the content of the frame (see p. 170) is reduced by the area occupied by the border, when a border is drawn around a frame. On the other hand, a block with a border needs more space within its superior layout object than the space required for the content of the block alone. Figure 32 shows an example for borders around frames and blocks. The direction of the layout path in this example should be noted since it determines the names of the edges.

Concatenation and layout object class

The attribute concatenation specifies whether the content of a basic logical object shall start in a new layout object or whether the content of several basic logical objects is to be connected during the layout process. The attribute layout object class indicates that the content of a basic logical object is to be laid out in a particular type of layout object.

The value of the attribute concatenation is either concatenated or non-concatenated:

concatenation = (concatenated | non-concatenated)

The attribute can be specified for layout styles and is used during the layout process when processing basic logical objects which refer to the layout style. Its value is determined according to the procedure described in Sect. 3.2.16; its default value is non-concatenated.

The specification of non-concatenated has the effect that the content associated with the basic logical object starts in a new basic layout object (a basic page or a block). In a document, for example, in which for every basic logical object a value of non-concatenated applies, the content of each basic logical object will be displayed in a basic layout object of its own.

If the value of the attribute is concatenated, the content associated with the basic logical object shall be connected with the content of that basic logical object which precedes it in the sequential order and shall be laid out in the same layout object. In other words, by specifying concatenated for the attribute concatenation the content of two (or more) basic logical objects can be combined to be treated as a single unit during the layout process. For the determination of the preceding logical object only those basic logical objects in the sequential order are taken into account which have the same attribute values for content architecture class, layout category and fill order as the logical object for which the value concatenated applies.

How the layout process shall be performed when the attribute has the value concatenated, and whether this attribute value is permitted or not, is is specified in Parts 6, 7 and 8 of the Standard where the content layout processes for the particular content architectures are described. At present, the value concatenated can only be used for character content, not for raster graphics or geometric graphics content.

A typical application for the attribute concatenation with the value concatenated is, for instance, the automatic numbering of chapter headings. Whereas the chapter number is created by means of the attribute content generator specified for a basic logical object, the actual text of the heading (without the number) is associated with another basic logical object (see the example on p. 104). In order to format the chapter number and the heading as a single unit the two basic logical objects have to be arranged in the corresponding sequential order and for the second object the attribute concatenation must have the value concatenated. (This attribute is not specified directly on the basic logical object but in a layout style to which the object refers.)

The value of the attribute layout object class is null or a *layout-object-class-id*:

layout object class = (null | *layout-object-class-id*)

The attribute can be specified for layout styles and is used during the layout process when processing logical objects which refer to the layout style. Its value is determined according to the procedure described in Sect. 3.2.16; its default value is null.

The specification of an identifier of a layout object class has the effect that the content associated with the object – in case of a composite logical object this is the content of all subordinate basic logical objects – is laid out within a single layout object belonging to this class. No other content of the document may be laid out within this layout object. (The *layout-object-class-id* may not refer to a block class.) The value null indicates that the layout of the content of the logical object is not restricted to a particular type of layout object, at least not by this attribute.

An example for the application of this attribute is the layout of different parts of a document into different page sets where for each page set a particular kind of page make-up may apply. In this case, each of these document parts would be combined to a separate composite logical object where each composite logical object would refer to a different page set by means of the attribute layout object class. (The attribute is not specified directly on the composite logical object but in a layout style to which the object refers.)

Balance and synchronization

The attributes balance and synchronization can be used to handle certain layout specifications for multi-column layout.

The value of the attribute balance is either null or a sequence of two or more *layout-object-id* or *layout-object-class-id*:

balance =
    (null | [[*layout-object-id* [*layout-object-id*]$^+$]]
    | [[*layout-object-class-id* [*layout-object-class-id*]$^+$]])

This attribute can be specified for composite layout objects and composite layout object classes without immediately subordinate blocks or block classes. For layout objects, only *layout-object-ids* are permitted, and for layout object classes only *layout-object-class-ids*. If the attribute specifies for a composite layout object class one or more *layout-object-class-ids*, the attribute generator for subordinates must also be present whose attribute value (a construction expression) must be able to generate the same sequence of layout object class identifiers as those specified by the attribute balance, i.e., the values of the attributes generator for subordinates and balance must be consistent. Furthermore, all the layout objects or layout object classes referenced by the attribute balance must have the same layout path (specified by the attribute layout path) and the same set of "permitted categories" (specified by the attribute permitted categories, see p. 128). Otherwise the attribute balance will be ignored.

If a value for the attribute is missing for objects of the specific structures a default value will be determined according to the procedure described in Sect. 3.2.16. If no explicit default value can be found within the document, the value null is assumed.

A value different from null specifies that the leading edges of the layout objects which are referenced by the *layout-object-ids* (or of the layout objects belonging to the referenced layout object classes) shall be aligned – as far as possible – along a line orthogonal to the direction of the layout path. The value null indicates that no alignment is required, at least not because of this attribute.

A typical application of this attribute is the alignment of text within a multi-column layout if all columns shall have the same height. In a two-column layout, for instance, the first column should not have the size of the whole page whereas the second column is considerably shorter; the layout process should rather produce the same height for both columns. Figure 33 gives an example how this could be achieved with the attribute balance.

object-class-id	generator for subordinates	balance
$[0\ 2\ 4]$	seq $[0\ 2\ 4\ 0]$ $[0\ 2\ 4\ 1]$	$[[0\ 2\ 4\ 0]\ [0\ 2\ 4\ 1]]$
$[0\ 2\ 4\ 0]$	. . .	null
$[0\ 2\ 4\ 1]$	. . .	null

Fig. 33: Alignment of column heights for multi-column layout

In this example, it is assumed that there exists a frame class within the generic layout structure with the *object-class-id* $[0\ 2\ 4]$. This frame class shall have have two subordinate frame classes with the *object-class-id* $[0\ 2\ 4\ 0]$ and $[0\ 2\ 4\ 1]$, specified by the attribute generator for subordinates. These two frame classes may represent two columns on a page into which text shall be laid out whereas the superior frame class may represent that area on the page where the two columns are put beside each other. In order to achieve columns of equal length, the attribute balance is specified for the superior frame class where the value of the attributes refers to the two subordinate frame classes.

The value of the attribute synchronization is either null, an object identifier expression or a *logical-object-id*:

synchronization =
    (null | '*object identifier expression*' | *logical-object-id*)

The attribute can be specified for layout styles and is used during the layout process when processing logical objects (except the document logical root) which refer to the layout style. Its value is determined according to the procedure described in Sect. 3.2.16; its default value is null.

If the layout style for which the attribute is specified is referenced from a logical object class, its value must be null or an object identifier expression; if the layout style is referenced from a logical object, its value must be null or a *logical-object-id.*

A value different from null indicates that two layout objects which contain the content of two logical objects shall be aligned along a line orthogonal to the layout path. The method of alignment is as follows: The trailing edge of the first block which contains content of the logical object referenced by the attribute, and the trailing edge of the first block which contains content of the logical object for which the attribute applies, are aligned along a line orthogonal to the layout path. The two blocks which are aligned must be placed in different frames. Furthermore, the fill order of the blocks (see p. 112) and the layout path within the frames must be the same. Otherwise, the attribute is ignored. The value null indicates that no alignment is required, at least not because of this attribute.

This attribute can be used, for instance, to align text and associated margin notes as shown in Fig. 34.

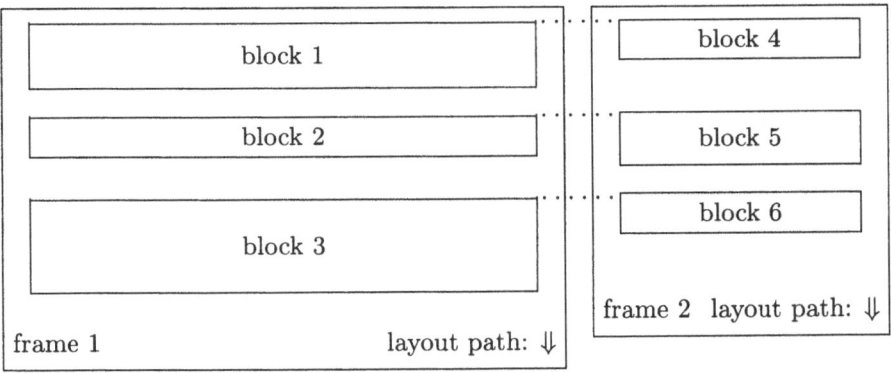

Fig. 34: Example for the attribute synchronization

In this example, the blocks 1 – 3 shall be paragraphs within a text whereas the blocks 4 – 6 shall be margin notes attached to the particular paragraphs. To align these blocks parallel to each other as indicated by the dotted lines, two frames are used, one containing the actual text and one the margin notes. The logical objects with the content of the margin notes refer to a layout style by means of the attribute layout style. This layout style specifies the attribute synchronization whose value is a reference to the logical object to which the margin note belongs.

Layout category, permitted categories, indivisibility, new layout object and same layout object

The remaining five attributes of this section – layout category, permitted categories, indivisibility, new layout object and same layout object – specify how the content of several logical objects shall be displayed in the specific layout structure.

The value of the attribute layout category is either null or a *layout-category-id*:

layout category $=$ (null | *layout-category-id*)

The attribute can be specified for layout styles and is used during the layout process when processing basic logical objects which refer to the layout style. Its value is determined according to the procedure described in Sect. 3.2.16; its default value is null.

This attribute assigns a so-called *layout category* to a basic logical object. (The value null indicates that the logical object has no layout category assigned.) The assignment of layout categories provides a mechanism to split the content of a document into several so-called *layout streams* (see p. 173) during the layout process. If a layout category is assigned to a basic logical object, the content associated with the object is only laid out in frames for which the attribute permitted categories specifies the layout category of the logical object. Within these frames, the content of the basic logical objects appears in the order given by the sequential order of the logical objects with the same layout category. An application for this attribute is given after the description of the following attribute permitted categories.

The value of the attribute permitted categories is either null or a set of *layout-category-ids*:

permitted categories $=$ (null | {$[layout\text{-}category\text{-}id]^+$})

This attribute can be specified for frames and frame classes whose immediately subordinate objects are blocks. If the attribute is missing for frames, a default value is derived according to the procedure described in Sect. 3.2.16. If no explicit value can be found, the value null is assumed.

If for a frame one or more *layout-category-ids* are specified by the attribute permitted categories, only the content of those logical objects which have one of these *layout-category-ids* assigned by means of the attribute layout category, is laid out within the frame. In other words, whenever a frame has a permitted category different from null, the layout process directs only the content of those layout streams into this frame which are explicitly specified. The value null indicates that the content which can be laid out in the frame is not restricted to logical objects which have a particular layout category.

A possible application for layout streams is, for instance, the automatic generation of a table of contents for a document. The table of contents shall be displayed on a page of its own after the layout process for the document has been carried out. Slightly simplified, this can be achieved by building the logical structure of the documents as outlined in Fig. 35.

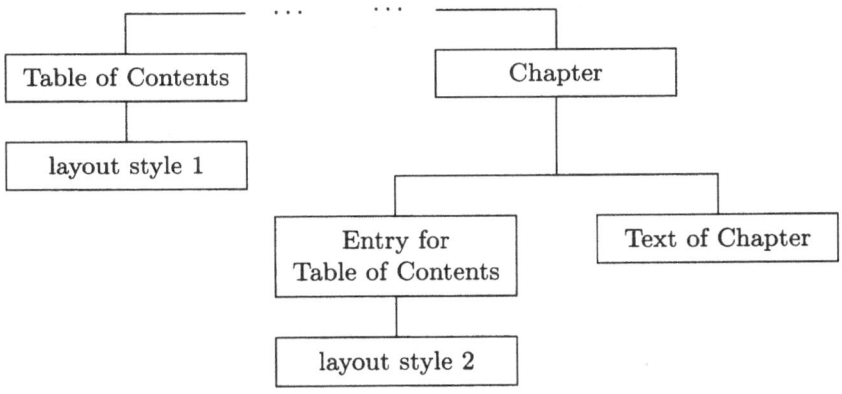

Fig. 35: Excerpt from the logical structure of a document

In this logical structure, there exists a basic logical object called "Table of Contents" with an associated layout style 1. This logical object may contain some text which is to be placed at the beginning of the table of contents, for example, a heading. Each logical object "Chapter" has a subordinate object called "Entry for Table of Contents" and a further subordinate object "Text of Chapter" whose internal structure is not important for this example. The "Entry for Table of Contents" has an associated layout style 2.

Both layout styles contain the attribute layout category where the value of the attribute layout category specifies the same *layout-category-id*, for instance,

layout category = table-of-contents

The generic layout structure contains a frame class which specifies this layout category as the only one whose content may be laid out in the corresponding frames by means of the attribute permitted categories:

permitted categories = {table-of-contents}

During the layout process the value of the attribute layout category creates a separate layout stream into which all content associated with "Table of Contents" and with the objects of type "Entry for Table of Contents" is directed. (In an actual document, there are usually several instances of this type of object, not only one as shown in the figure.) This content is laid out in a frame which specifies the assigned layout category as permitted. In other words, by means of the attribute layout category certain pieces of content of a document can be taken out of the "normal" processing order during the layout process and can be "redirected" to some particular place in the specific layout structure.

The value of the attribute indivisibility is either null, object type page, a *layout-object-class-id* or a *layout-category-id*:

indivisibility =
   (null | object type page | *layout-object-class-id* | *layout-category-id*)

The attribute can be specified for layout styles and is used during the layout process when processing logical objects (except the document logical root) which refer to the layout style. Its value is determined according to the procedure described in Sect. 3.2.16; its default value is null.

A value different from null specifies that the content associated with the logical object shall be laid out, if possible, within a single specified layout object. The layout object shall be a page for the value object type page, an instance of the specified layout class, if the value is a *layout-object-class-id*, or a frame for which the specified layout category is permitted. (The *layout-object-class-id* may not refer to a block class, but only to a page set class, page class or frame class.) The value null indicates that no constraint concerning the layout of the content of the logical object applies, at least not because of this attribute.

The application of this attribute – to keep certain parts of a document together and to prohibit column breaks or page breaks – should be obvious.

The value of the attribute new layout object is either null, object type page, a *layout-object-class-id* or a *layout-category-id*:

new layout object =
   (null | object type page | *layout-object-class-id* | *layout-category-id*)

The attribute can be specified for layout styles and is used during the layout process when processing logical objects (except the document logical root) which refer to the layout style. Its value is determined according to the procedure described in Sect. 3.2.16; its default value is null.

A value different from null specifies that the content associated with the logical object shall be laid out within a layout object of the specified kind which does not contain content from a preceding logical object. The layout object shall be a page for the value **object type page**, an instance of the specified layout class, if the value is a *layout-object-class-id*, or a frame for which the specified layout category is permitted. (The *layout-object-class-id* may not refer to a block class, but only to a page set class, page class or frame class.)

The value null indicates that no constraint concerning the layout of the content of the logical object applies, at least not because of this attribute.

As a typical application, for instance, this attribute could be used to start each chapter in a document on a new page. This can be be achieved by referring from each composite logical object which represents a particular chapter to a layout style which contains the attribute **new layout object** with the value **object type page**.

The value of the attribute **same layout object** consists of the two parameters **logical object** and **layout object**. The value of the parameter **logical object** is either null, a *logical-object-id* or an object identifier expression; the value of the parameter **layout object** is either **object type page**, a *layout-object-class-id* or a *layout-category-id*:

```
same layout object =
    {logical object = (null | logical-object-id |
                        ' object identifier expression ')
     layout object = (object type page | layout-object-class-id |
                        layout-category-id)}
```

The attribute can be specified for layout styles and is used during the layout process when processing logical objects (except the document logical root) which refer to the layout style. Its value is determined according to the procedure described in Sect. 3.2.16; its default value is the parameter **logical object** with the value null.

The value of the parameter **logical object** can only be a *logical-object-id* if the attribute is specified for a layout style which is referenced from a logical object. If the layout style is referenced from a logical object class, the value of the parameter must be an object identifier expression or null.

The value null indicates that this attribute has no effect on the layout process. In this case the parameter layout object may be missing; if a value is specified, it is ignored.

Otherwise, this attribute specifies that the content associated with two different logical object shall be laid out, if possible, within a single layout object as follows: The end of the content of the logical object which is referenced by the parameter logical object may be laid out in a basic layout object 1 (a block or basic page) and the start of the content of the logical object for which the attribute applies may be laid out in a basic layout object 2. These two basic layout objects shall then be subordinate to a single layout object of the type specified by the parameter layout object.

This attribute can be used, for instance, to avoid a page break or column break between a chapter heading and the first paragraph of the chapter. In this case, the attribute same layout object could be specified for the logical object representing the first paragraph. (The attribute is not specified directly on the logical object but on a layout style to which the logical object refers.) The parameter logical object would refer to the logical object representing the chapter heading, and the parameter layout object would specify the value object type page to avoid a page break, or the the parameter layout object would reference a frame representing the column in the case of a multicolumn layout.

In a similar manner, this attribute could be used to specify that a footnote text should start on the same page on which the reference to the footnote appears within the body text. If it is furthermore desired that the entire footnote text appears on the page, the attribute indivisibility with the value object type page must be specified for the logical object representing the footnote.

3.2.11   Attributes for Controlling the Imaging Process

Part 2 of the Standard describes five attributes for controlling the imaging process (see Sect. 3.6.3). These are the attributes colour, imaging order, medium type, page position and transparency. Further attributes for the imaging process are described in Parts 6, 7 and 8 where the different content architectures are specified.

The value of the attribute imaging order is a sequence of non-negative integers:

imaging order = $[\![\,[\,'non\text{-}negative\ integer\,']^+\,]\!]$

This attribute can be specified for composite pages and frames. It specifies the order according to which the immediate subordinate layout objects of a composite page or frame are handled by the imaging process. The order is specified similarly to the attribute subordinates: if for a composite page or frame the value of the attribute object identifier is $[\![x_1\ x_2\ \dots\ x_n]\!]$ and the value of the attribute imaging order is $[\![y_1\ y_2\ \dots\ y_m]\!]$, the subordinate layout objects are prepared for the presentation medium in the order $[\![x_1\ x_2\ \dots\ x_n\ y_1]\!]$, $[\![x_1\ x_2\ \dots\ x_n\ y_2]\!]$ ... $[\![x_1\ x_2\ \dots\ x_n\ y_m]\!]$.

If the attribute is missing, the subordinate layout objects are processed according to their sequential layout order. In other words, this attribute can be used to override the sequential layout order as given by the attribute subordinates during the imaging of layout objects. Different imaging orders may give different results, for instance, if the attribute transparency described hereafter applies.

The value of the attribute transparency is either transparent or opaque:

transparency = (transparent | opaque)

This attribute can be specified for pages, frames, blocks, page classes, frame classes, block classes and presentation styles. If it is missing for pages, frames or blocks, a default value is determined according to the procedure described in Sect. 3.2.16. If no explicit default value can be found, the value transparent is assumed.

The value of the attribute affects the imaging of intersecting layout objects on the presentation medium. If the value is opaque, the content of objects which precede the object for which this attribute value is specified are (totally or partially) invisible if they share the same area on the presentation medium. If the value is transparent, the content of earlier imaged layout objects remains visible. The order according to which the layout objects are imaged depends on the sequential layout order (see p. 91) and on the value of the attribute imaging order, if specified.

Figure 36 shows an example for the application of the attribute transparency. The borders in this figure show the areas on the presentation medium which belong to each particular layout object.

Of course, this is example is not a very useful application of the attribute transparency. A more realistic application for a transparent overlay of layout objects are form sheets where the predefined text is contained in one layout object and the text which is to be filled into the fields of the form sheet, is contained in one or more other layout objects which are laid over the first layout object in a transparent manner.

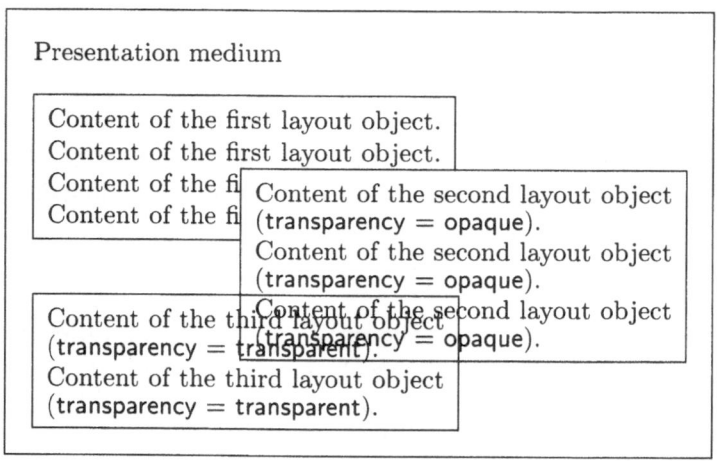

Fig. 36: Example of the attribute transparency

The value of the attribute colour is either colourless or white:

colour = (colourless | white)

This attribute can be specified for pages, frames, blocks, page classes, frame classes, block classes and presentation styles. If it is missing for pages, frames or blocks, a default value is determined according to the procedure described in Sect. 3.2.16. If no explicit default value can be found, the value colourless is assumed.

The value of this attribute specifies the background colour of the area associated with the layout object. The value white is only permitted if the value of the attribute transparency is opaque. Further explanations on the use of the attribute are given in Sect. 3.6.3.

The value of the attribute page position consists of two non-negative integers:

page position = ⟦ ' non-negative integer ' ' non-negative integer ' ⟧

This attribute can be specified for pages and page classes. If it is missing for a page, a default value is determined according to the procedure described in Sect. 3.2.16. If no explicit default value can be found, the two integers are determined according to the procedure described on p. 178.

The first integer specifies the horizontal, the second integer the vertical distance from the upper left corner of the presentation medium to the upper left corner of the layout object page. (It should be noted that in ODA terminology the term "page" denotes a layout object and does not have the usual meaning of a page as a physical object, i.e., a sheet of paper.) The distances are measured in scaled measurement units (see p. 148).

The value of the attribute medium type consists of the two (optional) parameters nominal page size and side of sheet. The value of the parameter side of sheet is a pair of positive integers, the value of the parameter side of sheet is either recto, verso or unspecified:

medium type =
    {[nominal page size = [['positive integer' 'positive integer']]
    [side of sheet = (recto | verso | unspecified)] }

This attribute can be specified for pages and page classes. The parameter nominal page size defines the size of the area of the presentation medium where the first integer is the width and the second integer the height. The dimensions are given in scaled measurement units. The second parameter specifies whether the page is to be printed on the recto or verso side of a sheet of paper or whether this is not important (unspecified).

This parameter is, if course, meaningless when the page is imaged on a computer screen. When using a printer which can print on both sides of a sheet of paper this parameter can be used, for instance, to print odd-numbered pages on verso and even-numbered pages on recto sides.

If this attribute or one of its parameters is missing for pages, a default value is determined according to the procedure described in Sect. 3.2.16. If no explicit default value can be found, the size of an ISO A4 page is assumed for the parameter nominal page size, i.e., a width of 210 mm = 9920 basic measurement units and a height of 297 mm = 14030 basic measurement units (see p. 177). The default value for the parameter side of sheet is unspecified.

3.2.12 Security Attributes

The attributes enciphered, sealed, sealed document profile information and enciphered information were introduced by the Addendum on Security and are used to handle certain security aspects when interchanging ODA documents (see Sect. 3.5).

The value of the attribute enciphered consists of the two parameters
enciphered subordinates and protected part identifier. The value of the
parameter enciphered subordinates is either none, all or the sub-parameter
partial whose value is a sequence of one or more non-negative integers.
The value of the parameter protected part identifier is a *protected-part-id*
(see p. 73):

enciphered =
   {enciphered subordinates = (none | all |
                            partial = [[' *non-negative integer* ']+])
     protected part identifier = *protected-part-id*}

This attribute can be specified for basic object classes, basic objects
and composite objects. It specifies whether or not subordinate objects –
including their associated content portions – of the object for which the
attribute is specified, are enciphered and, if so, the "name" of the enciphered document body part.

The value none for the parameter enciphered subordinates indicates that
the subordinate objects are not enciphered. In this case – and in this case
only – the parameter protected part identifier is not specified. The value all
indicates that all subordinate objects, if any, and their associated content
portions are enciphered. If the sub-parameter partial with a sequence of
integers is given for the parameter enciphered subordinates – this is only
permitted for basic objects in a formatted processable document – only
the content portions identified by these integers are enciphered. The
integers in the sequence must also appear in the attribute content portions
specified for the object (see p. 95).

If the value of the parameter enciphered subordinates is different from
none, the parameter protected part identifier identifies the pre-enciphered
or post-enciphered document body part containing the enciphered subordinate objects, if any, and content portions.

The value of the attribute sealed consists of the two parameters sealed
status and seal identifiers. The value of the parameter sealed status is
either yes or no. The value of the parameter seal identifiers is a set of seal
identifiers:

sealed =
   {sealed status = (yes | no)
   seal identifiers = {[' *seal identifier* ']+}}

This attribute can be specified for all objects, object classes, layout
styles and presentation styles. If the parameter sealed status has the

value yes, the parameter seal identifiers specifies a list of seal identifiers in which the constituent is incorporated. These seal identifiers appear in the document profile attributes pre-sealed document body parts or post-sealed document body parts as values of the parameters seal identifier. If the parameter sealed status has the value no, the parameter seal identifiers is not specified. If the attribute is missing, .the parameter sealed status has the default value no.

The value of the attribute sealed document profile information is a set of attributes from the document profile (see Sect. 4.1):

sealed document profile information = {'*document profile attributes*'}

This attribute can be specified for sealed document profiles of the protected part. It lists those attributes of the document profile for which a seal is given. The value null may be given for each attribute to indicate that this attribute is sealed as absent in the document profile.

The value of the attribute enciphered information is an *octet string*:

enciphered information = '*octet string*'

This attribute can be specified for constituents of the protected part and represent the enciphered information of these constituents. An authorized receiver of a document has to decipher these attribute values to reconstruct the unenciphered version of the document before it can be processed (see Sect. 3.5). In particular, the attributes of the document profile are determined by deciphering this attribute for an enciphered document profile; by deciphering this attribute for an pre-enciphered or post-enciphered document body part those constituents of the document are reconstructed that have been enciphered.

3.2.13   Other Attributes

There are two other attributes left which have not yet been discussed in the previous sections: protection and default value lists.

The value of the attribute protection is either protected or unprotected:

protection = (protected | unprotected)

This attribute can be specified for all objects and object classes of the logical structure. If it is missing for an object of the specific logical structure, a default value is determined according to the procedure described in Sect. 3.2.16. If no explicit default value can be found, the value **unprotected** is assumed.

The value **protected** indicates that the receiver of a document is not allowed to modify any attributes of the logical object for which this value is specified. In the case of basic logical objects, the receiver is also not allowed to modify content portions which belong to the logical object. If the value **unprotected** is specified, the receiver may modify the logical object and associated content portions. This attribute can be used to assign a kind of "write protection" to certain parts of a document.

The value of the attribute **default value lists** is a set whose elements are again sets. The elements of these subordinate sets consists each of an attribute name and an attribute value:

$$\textsf{default value lists} = \left\{ \left[\left\{ \left[\text{'} \textit{attribute name} \text{'} \ \text{'} \textit{attribute value} \text{'} \right]^+ \right\} \right]^+ \right\}$$

This attribute can be specified for composite objects and composite object classes. Each pair, an attribute name and an attribute value, defines a default value for the respective attribute which may be taken into account during the procedure for determining the default value of defaultable attributes (see Sect. 3.2.16).

The concept behind the attribute **default value lists** is the following: Each composite logical object or composite logical object class can specify a list of default values for its subordinate objects or object classes by means of this attribute. A document logical root class, for instance, can specify two lists (sets): one for composite logical objects and one for basic logical objects.

Only one list is permitted for each type of subordinate object. For example, it is not allowed to specify for a page two lists of default values for attributes of frames, even if more than one frame is subordinate to the page. Furthermore, only one default value may appear in a list for each particular attribute. There exists one exception to this rule: more than one default value may be specified for the attributes **presentation style** and **presentation attributes**, if the values of these attributes correspond to different content architecture classes, i.e., the pairs "**presentation style** *attribute value*" and "**presentation attributes** *attribute value*" may appear for character content, raster graphics content and geometrics graphics content within one list.

The attributes for which default values may be specified by means of the attribute **default value lists**, depend on the type of object as shown in Fig. 37.

Type of object	Permitted attributes in default value list	
page	content architecture class*	colour
	dimensions	medium type
	page position	presentation style*
	presentation attributes*	sealed
	transparency	
frame	border	colour
	dimensions	layout path
	permitted categories	position
	sealed	transparency
block	border	colour
	content architecture class	dimensions
	position	presentation style
	presentation attributes*	sealed
	transparency	
composite logical object	layout style	protection
	sealed	
basic logical object	content architecture class	layout style
	presentation style	protection
	sealed	
*: only for basic pages		

Fig. 37: Permitted attributes in the value of the attribute default value lists

The table in Fig. 37 is to be read, for instance, as follows: in a list containing default values for objects of type "page", only default values for the attributes content architecture class, colour, dimensions, medium type, page position, presentation style, presentation attributes and transparency are permitted. Such a list may appear in the value of the attribute default value lists for objects or object classes at a hierarchical higher level, i.e., at a page set or document layout root level.

It is impossible to specify default values for attributes of the document logical root or document layout root since these objects do not have superior objects. Furthermore, the specification of default values for attributes of page sets is not allowed.

3.2.14   Alternative Descriptions

A basic logical object or basic layout object (basic page or block) may be represented by its so-called *primary description* and one or more so-called *alternative descriptions*. (The concept of primary and alternative descriptions was introduced by the Addendum on Alternate Representations; it was not contained in the initially published version of the Standard.) Both a primary description and an alternative description is the basic object itself with possibly associated content portions.

For a basic object representing an alternative description, i.e., for a basic object referenced from another basic object by means of the attribute alternative, the attribute primary – indicating the associated primary description – must be specified. In other words, by the absence or presence of the attribute primary it can be distinguished whether a basic object is a primary or an alternative description. A basic object representing an alternative description may again specify another basic object as an alternative description, i.e., the alternative descriptions may build a chain as shown in Fig. 38.

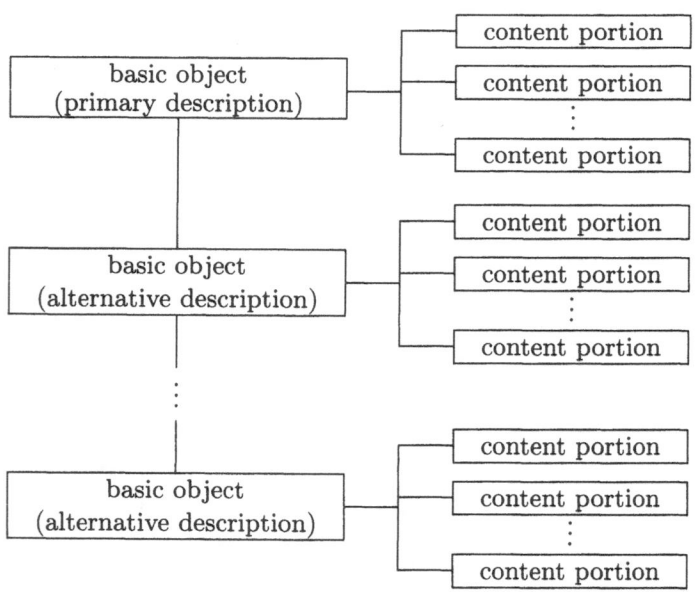

Fig. 38: Primary and alternative descriptions

The concept of alternative descriptions was introduced for the following reasons. The originator of an ODA document may want to send the document to recipients where some of them may not be able to process the document completely because a recipient's system may only have implemented a subset of the Standard. For instance, a recipient may not be able to process all presentation attributes of a particular content architecture or such a system may only process raster graphics content but not geometric graphics content.

The originator may therefore specify one or more alternative descriptions of a basic object in addition to the primary description which usually represents the intent of the originator most appropriately. For example, the originator may transform a geometric graphics picture into a raster graphics image and include the raster image as an alternative representation of the primary geometric picture. Obviously, such a method is more powerful than using the attribute alternative representation (see p. 100) which would only provide a very elementary "fallback" mechanism.

The order of the primary and alternative descriptions forms a hierarchy, i.e., as soon as a recipient's system encounters a description it can process, this basic object and its associated content portions, if present, shall be used in the document; all other (primary or alternative) descriptions are to be ignored. In other words, even if several descriptions exist in a received document, only one description is selected for each basic object before the document is processed.

The ODA Standard does not specify explicitly at which stage the selection process has to take place. This may be when decoding the ODIF data stream, when the layout process starts or – in case of a formatted document – when the imaging process begins. In particular, the concept of primary and alternative descriptions does not introduce additional complexities for the layout process or imaging process since they are only concerned with the single description selected for each particular basic object.

3.2.15   Complete Generator Sets, Partial Generator Sets and Factor Sets

When describing the constituents of ODA documents the terms *complete generator set*, *partial generator set* and *factor set* have been used at several places. The meaning of these terms is described in this section.

When introducing the generic structures of the ODA architectural model in Sect. 2.1.2 it was assumed that the generic structure of an ODA document describes, as far as possible, the associated specific structure. This would mean, for instance, that all objects in the specific structure have associated object classes in the generic structures and all "construction rules" for the specific structures can be found in the generic structure. However, this is not the case for all ODA documents, due to the fact that generic structures in ODA documents may serve two different purposes:

- As one purpose, generic structures may be used for "consistency checks" between a particular document and its associated document class. In this case all "construction rules" for the specific structures must indeed be specified in the generic structures.

  As an example, consider the generic logical structure of the business letter shown in Fig. 4. If a letter were to contain a "postscript" besides the "greetings", "signature", "name" and "enclosure", its specific logical structure would be inconsistent with this generic logical structure, i.e., such a letter would not belong to the document class described by this generic structure.

- As a second purpose, generic structures in ODA documents are also used to "factorize" certain pieces of information from the specific structures into the generic structures, for instance, to reduce the amount of information which has to be stored or interchanged electronically. (The salutation "Best Regards", for example, might be factorized into a generic content portion associated with the logical object "greetings", so avoiding an inclusion of this text into each particular letter belonging to this document class.

  For this purpose, it is rather obvious that a weaker degree of correspondence between generic structures and specific structures would be necessary. Even a letter with a "postscript", for instance, may be considered consistent with the generic structure of Fig. 4: it is only requested that the specific structure agrees with the generic structure "in principle" but it is permitted that the specific structure contains objects which are not reflected in the generic structure.

A look at the generic structure of a document is usually not sufficient to decide its role in a particular document. This is explicitly specified in the document profile by the attributes **generic layout structure** and **generic logical structure** whose value is either **complete generator set, partial generator set** or **factor set** (see p. 186). The values of these attributes define whether a generic layout structure or generic logical structure, if present, shall be considered as a so-called *complete generator set, partial generator set* or *factor set.* This is important for certain other attributes in the document, for example, whether or not the attribute **object class** is a mandatory attribute (see Note 2 on p. 45).

The rules which apply to these three kinds of generic structures and which are slightly different for generic logical and generic layout structures are explained in the next two sections.

Generic logical structures

If the generic logical structure is a complete generator set, i.e., if the attribute **generic logical structure** of the document profile has the value **complete generator set**, each object in the specific logical structure has to refer to an object class in the generic logical structure by means of the attribute **object class**. The subordinate objects of a particular constituent in the specific structure must be consistent with the rules specified by the associated object class, i.e., the value of the attribute **subordinates**, specified for the constituent, must be contained within the value range of the attribute **generator for subordinates**, specified for the object class.

If the generic logical structure is a factor set, i.e., if the attribute **generic logical structure** of the document profile has the value **factor set**, the attribute **generator for subordinates** may not be specified for any object class in the generic logical structure. In other words, all object classes are "isolated" and no hierarchical relations between the object classes are specified. This means in particular, that no rules are defined for the hierarchical tree structure of the objects in the specific logical structure. The objects in the specific logical structure may – but need not necessarily – refer to an object class by means of the attribute **object class**.

If the attribute **generic logical structure** of the document profile has the value **partial generator set**, this indicates that the object classes of the generic layout structure may specify the attribute **generator for subordinates**, thus giving construction rules for the specific logical structure, but this structure is not necessarily completely controlled by this attribute.

Generic layout structures

If the generic layout structure is a complete generator set, i.e., if the attribute **generic layout structure** of the document profile has the value **complete generator set**, each composite object in the specific layout structure has to refer to an object class in the generic layout structure by means of the attribute **object class**. For basic layout objects, i.e., for blocks and basic pages, such a reference is permitted, but not necessarily required. The subordinate objects of a particular constituent in the specific layout structure must be consistent with the rules specified by the associated object class, i.e., the value of the attribute **subordinates**, specified for the constituent, must be contained within the value range of the attribute **generator for subordinates**, specified for the object class.

Concerning factor sets and partial generator sets, the same rules as for the logical structures apply: if the generic layout structure is a factor set, the attribute generator for subordinates is not permitted for layout object classes; if the generic layout structure is a partial generator set, this attribute is permitted but need not necessarily control the specific layout structure completely.

3.2.16   Determination of Attribute Values for Defaultable Attributes

Many ODA attributes are classified as defaultable: if the value of such an attribute is not specified directly for a particular object, a so-called *default value* will be determined. This is performed whenever an attribute value is needed during the processing of the object, depending on the semantics of the attribute. For example, the value of the attribute protection of a particular object will be evaluated when the object shall be edited, the value of the attribute layout path will be determined during the layout process, and the value of the attribute imaging order will be evaluated during the imaging process.

Concerning the determination of the value of a defaultable attribute, the following nine steps may be possible:

a) The value of the attribute is specified explicitly on the object.

b) The value of the attribute is specified on a layout style or presentation style which is referenced by the object by means of the attribute layout style or presentation style, respectively. The path for finding the attribute value is therefore: object → layout style → attribute value or object → presentation style → attribute value.

c) The value of the attribute is specified on an object class which is referenced by the object by means of the attribute object class. The path for finding the attribute value is therefore: object → object class → attribute value.

d) The value of the attribute is specified on a layout style or presentation style which is referenced by an object class by means of the attribute layout style or presentation style, respectively. The object class, in turn, is referenced from the object by means of the attribute object class. The path for finding the attribute value is therefore: object → object class → layout style → attribute value or object → object class → presentation style → attribute value, respectively.

e) The value of the attribute is specified on an object class in a resource document which is referenced by an object class associated with the object by means of the attribute resource. The path for finding the attribute value is therefore: object → object class → object class in resource document → attribute value.

f)   The value of the attribute is specified on a layout style or presentation style which is referenced by an object class in a resource document by means of the attribute layout style or presentation style, respectively. The object class in the resource document is referenced by an object class associated with the object by means of the attribute resource. The path for finding the attribute value is therefore: object → object class → object class in resource document → layout style → attribute value or object → object class → object class in resource document → presentation style → attribute value, respectively.

g)   The value of the attribute is specified by the attribute default value lists which is specified for an object at a higher level within the hierarchical structure. If an attribute value can be found in several default value lists, the value derived from the lowest hierarchical level in the structure is used. At each level, default value lists in object classes or object classes in resource documents which may be associated with an object are also taken into account. Furthermore, if any of these default value lists specifies a layout style or presentation style, it is examined whether the style specifies a value for the attribute. The paths for finding the attribute value are therefore: object → superior object → default value list → attribute value, object → superior object → default value list → layout style → attribute value, object → superior object → default value list → presentation style → attribute value, object → superior object → object class → default value list → attribute value, object → superior object → object class → default value list → layout style → attribute value, object → superior object → object class → default value list → presentation style → attribute value, object → superior object → object class → object class in resource document → default value list → attribute value, object → superior object → object class → object class in resource document → default value list → layout style → attribute value, or object → superior object → object class → object class in resource document → default value list → presentation style → attribute value.

h)   If no attribute value can be found according to the steps a)–g), it is examined whether an attribute value is specified by means of the attribute document application profile defaults in the document profile. The path for finding the attribute value is therefore: object → document profile → attribute value.

i) If no attribute value can be found according to the steps a)–h), the default value explicitly defined in the ODA Standard is taken. These are the values which have been given in Sects. 3.2.5 – 3.2.13 for each defaultable attribute.

As mentioned above, the values of defaultable attributes are determined during the processing of ODA documents. Since the ODA Standard describes three different processes, the editing, layout and imaging process, the time of determination depends on the kind of process for which a certain attribute value is needed.

Most attributes influence the layout process, i.e., their value is determined during the layout process. A few attributes are important for the imaging process (see Sect. 3.6.3) and essentially only one attribute, namely **protection**, is required during the editing process.

Attribute name	Steps	Attribute name	Steps
application comments	a c e i	new layout object	b d f g i
balance	a c e i	object type	a c
bindings	a c e i	offset	b d f g i
block alignment	b d f g h i	page position	a c e g h i
border	a – i	permitted categories	a c e g i
coding attributes	a c h^2	position	a c e g i
colour	a – i	presentation attributes	a – g h^2 i
concatenation	b d f g i	protection	a c e g i
content architecture class	a c e g h i	same layout object	b d f g i
dimensions	a c e g h^1 i	sealed	a – i
fill order	b d f g i	separation	b d f g i
indivisibility	b d f g i	synchronization	b d f g i
layout category	b d f g i	transparency	a – i
layout object class	b d f g i	type of coding	a c h
layout path	a c e g h i	user-readable comments	a c e i
medium type	a c e g h i	user-visible name	a c e i

1: Step is only performed if the attribute value is to be determined for a page.

2: The permitted coding attributes and presentation attributes for which a value can be given by means of the attribute **document application profile defaults** depend on the content architectures.

Fig. 39: List of the defaultable attributes and the steps applied for the determination of their value

Not all of the steps a)–i) described above are performed for each attribute: the applicable steps depend on the attribute. Figure 39 lists for each defaultable attribute the steps which may be carried out to determine its value.

Furthermore, it should be noted that several attributes have parameters whose values are independently defaultable. In such a case the parameter values are determined according to the same steps which apply for the attribute value as a whole. Consider, for instance, the attribute medium type which has the independently defaultable parameters nominal page size and side of sheet. If the parameter side of sheet is missing for the attribute, for example, and has to be determined this is done according to the steps c), e), g), h), and i) until a value for the parameter has been found.

3.2.17   Determination of Values for Attributes of Styles

An ODA document may contain root styles and derived styles. To determine the value of an attribute in a layout style or presentation style the following three steps are performed.

a)   If the attribute is specified for a style, its value is used.

b)   If the style is a derived style – i.e., if the attribute derived from is specified for the style – and the attribute is specified in the style referenced by the attribute derived from, this attribute value is used. This step, i.e., examining the referenced styles of derived styles, may be repeated until the root style is reached.

c)   If the root style is reached and no attribute value has been found, the default value for the attribute is determined according to the procedures described in the preceding section, if the attribute is classified as defaultable. Otherwise, the value of the attribute is not applied.

As soon as an attribute value has been determined no additional steps according to this procedure are carried out.

3.2.18   Measurement Units

For several attributes or parameters, for instance, for the parameter nominal page size of the attribute medium type, their values represent physical sizes. These sizes can be specified in two different ways in the ODA Standard: as *basic measurement units* and as *scaled measurement units*.

A basic measurement unit is an absolute unit and corresponds to $\frac{1}{1200}$ inch which is about 0.021167 mm.

A scaled measurement unit is a relative unit whose absolute value can be computed by means of the attribute unit scaling (see p. 205) in the document profile. If this attribute is not specified in the document profile, one scaled measurement unit corresponds to one basic measurement unit.

3.3   Components of ODA Documents

As mentioned already at several places above, different types of constituents may appear in an ODA document: logical objects, logical object classes, layout objects, layout object classes, layout styles, presentation styles, content portions and a document profile. The ODA Standard defines a set of rules which types of constituents are permitted in a particular document and which types have to be present. These rules are explained in this section.

3.3.1   Document Architecture Classes

The ODA Standard defines three so-called *document architecture classes*, namely the *formatted document architecture class*, the *processable document architecture class* and the *formatted processable document architecture class*.

A document belonging to the formatted document architecture class is intended to be displayed on a presentation medium, i.e., only the imaging process can be applied to such a document. A modification of the content or the layout of the document is impossible. A document is usually transmitted in this form, if the receiver is not expected to make any modifications to it.

If a document belongs to the processable document architecture class, both its content and its layout can be modified. Such a document contains no specific layout structure, i.e., before such a document can be displayed on a presentation medium a layout process has to been carried out.

A document belonging to the formatted processable document architecture class is a mixture of the two classes above. On the one hand, an imaging process can be immediately applied to the document, i.e., a receiver can display the document on a presentation medium in exactly the same way as it was intended by the originator. On the other hand, both the content and the layout of the document can still be modified;

if this is done a layout process has to been carried out, of course, before the modified document can be displayed.

Each ODA document belongs to exactly one of these classes. Depending on the class to which a document belongs, different rules determine which types of constituents are required, permitted or prohibited in the document. These rules are as follows:

a) A document belonging to the formatted document architecture class always contains
    – a document profile and
    – a specific layout structure.
    Optionally, the following types of constituents may be present:
    – a generic layout structure and
    – presentation styles.

b) A document belonging to the processable document architecture class always contains
    – a document profile and
    – a specific logical structure.
    Optionally, the following types of constituents may be present:
    – a generic logical structure,
    – a generic layout structure,
    – layout styles and
    – presentation styles.

c) A document belonging to the formatted processable document architecture class always contains
    – a document profile,
    – a specific logical structure,
    – a specific layout structure and
    – a generic layout structure.
    Optionally, the following types of constituents may be present:
    – a generic logical structure,
    – layout styles and
    – presentation styles.

It should be noted that the generic structures and the styles need not necessarily be directly contained within the document body but they may be stored separately within so-called *generic documents* (see Sect. 3.4.1).

A generic document is assigned to one of the document architecture classes according to the following rules:

a) If a generic document contains layout object classes but no logical object classes, it belongs to the formatted document architecture class.

b) If a generic document contains logical object classes but no layout object classes, it belongs to the processable document architecture class.

c) If a generic document contains both layout object classes and logical object classes, it belongs to the formatted processable document architecture class.

By these rules each particular ODA document is assigned to one of the document architecture classes. This assignment is also explicitly expressed by the value of the attribute document architecture class which has to be specified in the document profile of each document.

3.3.2   Documents Containing Only a Document Profile

In addition to the rules given in the preceding section, the ODA Standards allows that a document may consist of the document profile alone, though this is not a document in the usual sense: it contains no content and, as a consequence, cannot be displayed on a presentation medium.

Nevertheless, it may be sensible to transmit such a "document" to a receiver since the document profile contains a lot of information about the document. For instance, the document profile specifies the kind of content – character content, raster graphics or geometric graphics – which is contained in the document. This information may be sufficient for the receiver to decide whether or not he would be able to process the document. If, for example, the document profile specifies that raster graphics content appears in the document, a transmission of the complete document would be useless for a receiver who cannot process raster graphics.

To each document consisting of a document profile alone belongs always a "real" document with the usual constituents such as objects and object classes. Splitting the document profile from a particular document and creating a new ODA document out of it which is electronically interchanged, is only done to avoid unnecessary transfers of documents to receivers who cannot process all valid ODA documents (see also Sect. 2.2).

3.3.3   Simple-Structured CCITT Documents

For reasons of compatibility with existing CCITT Recommendations, a certain type of document has been included in the the ODA Standard for which the "normal" specifications, for instance, on the usage of attributes, have been slightly modified. These documents shall be called *simple-*

structured CCITT documents. (This term is not used within the Standard itself.)

Only the following constituents are allowed within such documents:

- a document profile,
- layout object classes of type document layout root, basic page and block,
- layout objects of type document layout root, basic page and block,
- presentation styles and
- content portions.

Logical objects and object classes, layout styles and layout objects or object classes of type composite page or frame are not allowed. The following exceptions from the normal rules for ODA documents are permitted:

- The attribute **object identifier** may be missing for layout objects.
- The attribute **subordinates** may be missing for layout objects.
- The attribute **content portions** may be missing for basic layout objects.
- The attribute **content identifier layout** may be missing for content portions. (The attribute **content identifier logical** cannot be specified since such documents contain no logical structure and, therefore, no content portions can be associated with logical objects or object classes.)

It should be noted that many attributes cannot be used in these simple-structured CCITT documents, namely all those attributes which only appear in logical structures or layout styles.

Since the attributes **object identifier**, **content identifier layout**, **subordinates** and **content portions** may be missing in such documents, the hierarchical structure of the objects – for instance, to which page a particular block is subordinate or to which block a particular content portion belongs – would not be visible without additional provisions. Its is therefore specified that these documents can only be interchanged using the so-called interchange format class B (see p. 213). This interchange format allows the determination of the hierarchical structure of the constituents in the document only by inspection of the sequential order of the constituents in the data stream. The above mentioned four attributes can therefore be missing without loss of information.

3.4   Storage and Interchange of ODA Documents

It is usually desirable to keep the amount of data which is needed for the storage or interchange of ODA documents as low as possible. For this purpose, the concept of *generic documents* was included in the ODA Standard where a generic document is either a so-called *external document class description* or a *resource document*. These two kinds of generic documents can be referenced from a particular document by means of the attributes external document class or resource document which can be specified in the document profile.

3.4.1   Generic Documents

A generic document.is an ODA document whose content is as described in the following three cases:

a) – A document profile,
    – logical object classes which build either a complete generator set, partial generator set or factor set (see Sect. 3.2.15),
    – optionally layout styles,
    – optionally presentation styles and
    – optionally generic content portions;
b) – a document profile,
    – layout object classes which build either a complete generator set, partial generator set or factor set,
    – optionally presentation styles and
    – optionally generic content portions;
c) – a document profile,
    – logical object classes which build either a complete generator set, partial generator set or factor set,
    – layout object classes which build either a complete generator set, partial generator set or factor set,
    – optionally layout styles,
    – optionally presentation styles and
    – optionally generic content portions.

Concerning the storage and interchange, a generic document is considered a separate ODA document. However, the processing of generic documents alone is not sensible. In particular, since these documents are lacking any specific logical or specific layout structure, a layout process or imaging process cannot be applied to them. Such documents are not "displayable"; they only contain descriptions of properties of those documents which refer to them.

A generic document can be referenced in two ways: it is either referenced by a particular document as an external document class description (by means of the attribute external document class in the document profile) or as a resource document (by means of the attribute resource document in the document profile).

3.4.2 External Document Class Descriptions

If a particular document refers to a generic document as an *external document class description* by specifying the attribute external document class in the document profile, the following rules apply.

- If the document itself contains no generic structures, the generic structures of the generic document are used as the generic structures of the referencing document. This requires, of course, that the generic structures are consistent with the specific structures. Styles may be present in both the generic document and the referencing document. In particular, a constituent in the referencing document may refer to a style in the generic document.

- If the document itself contains a generic logical or generic layout structures, the generic structures of the generic document are ignored. Styles, however, may still be "imported" into the referencing document.

- If a layout style or a presentation style in the generic document has the same *layout-style-id* or *presentation-style-id* as a layout style or presentation style in the referencing document, the style in the generic document is ignored.

In other words, the generic structures in a generic document are imported on an "all-or-nothing" basis, whereas the styles in a generic document may be used in addition to the styles in the referencing document.

It should be noted, that a particular document which refers to a generic document as an external document class description, can in practice only be processed if all information in the generic document is taken into account. For instance, if an object refers to an object class in the generic document by means of the attribute object class, the precise *object-class-id* must be known (see p. 96).

The main purposes of external document class descriptions is the separate storage, and maybe interchange, of generic and specific structures. This may be useful, if several documents use identical generic but different specific structures.

3.4.3   Resource Documents

A particular document may refer to a generic document as a *resource document* by specifying the attribute **resource document** in its document profile.

In contrast to an external document class description, the object classes in a resource document can be used in addition to the object classes in the referencing document. The references to the object classes in a resource document are made by the attribute **resource** which can be specified for all object classes in a referencing document. In particular, attributes which are usually specified for object classes need not be specified in the object classes of a given document but may be "factorized" into object classes in a resource document.

The reference to an object class in a resource document is not done directly by specifying the *object-class-id* of the object class, but a "symbolic name" is used for identifying these object classes (see the description of the attribute **resources** on p. 188). This provides a certain degree of independence of the *object-class-ids* in a resource document and of the *object-class-ids* used in documents referring to the resource document and allows a more flexible use of resource documents in different applications.

The concept of resource documents can be used to extend the concept of document classes to a certain degree. For instance, a particular organization may have a set of different ODA documents classes for their different types of documents. Certain properties of all these document classes, however, might be identical, for example the height and width of the pages. Such common properties could then be stored in resource documents which are referenced by all document classes.

3.5   Security Aspects

The ODA Standard was developed for the interchange of documents in an open systems environment, i.e., a document transmitted by an originator will usually travel through an electronic network before it finally reaches the intended receiver. The originator and receiver will usually not know the intermediate places – such as mail servers – which the document passes. In particular, an unauthorized receiver may get access to the document on its way through a network as schematically shown in Fig. 40.

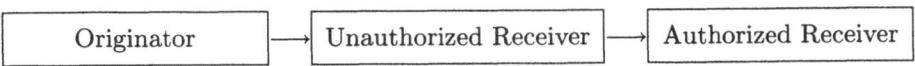

Fig. 40: Access to a document by authorized and unauthorized receivers

This problem immediately raises a number of questions relating to security aspects and, therefore, the Addendum on Security was developed. This Addendum addresses the following issues:

1. Confidentiality

   Confidentiality means that an unauthorized receiver is not able to get a semantic knowledge of specified parts of a document, i.e., slightly simplified, an unauthorized receiver cannot "read" the specified parts. Confidentiality is achieved by applying an encryption technique on such parts: the originator enciphers parts of the document and only an authorized receiver can decipher them again.

2. Integrity

   Integrity means that an authorized receiver can verify that a document or particular parts of it have not been changed after the originator initiated the transmission of the document. In particular, an authorized receiver can verify that an unauthorized receiver did not make any changes to the document and, also, that no transmission errors occurred. To be able to perform an integrity test on the receiver's side, the originator adds one or more so-called *seals* to the document.

3. Authenticity

   Authenticity means that an authorized receiver can identify the originator of the document. In particular, an originator cannot send a document to a receiver and pretend to be someone else; the receiver could detect such a fake. To be able to perform an authenticity test on the receiver's side, the originator adds a seal to the document.

4. Non-repudiation of origin

   Non-repudiation of origin means that a person cannot deny being the originator of a document. In particular, the receiver of a document can prove to a third party, for instance, in a legal case, that a particular person was, in fact, the originator of the document.

Non-repudiation of origin can also be achieved by adding a seal to a document.

5.   Different groups of authorized receivers

A document is often transmitted not only to a single authorized receiver but to a group of authorized receivers where different security aspects may apply to each member of the group. For example, the originator may provide access to a particular part of the document to receiver X but not to receiver Y. The Addendum on Security solves this problem by allowing different encryption techniques and seals within one document where a particular encryption technique and seal can be handled exclusively by a certain authorized receiver.

Of course, to address these security aspects requires additional agreements between an originator and an authorized receiver. In particular, originator and receiver must establish a security policy. For instance, they may agree that each interchanged document is enciphered and contains seals to allow the verification of integrity and authenticity; any document without such seals would be ignored by the recipient.

The Addendum on Security neither defines a particular encryption technique nor specifies algorithms how seals are to be created.

3.5.1   Encipherment

State-of-the-art methods for the encipherment of information are often based on so-called *keys*. The originator uses a certain algorithm for encipherment making use of a key which is often a very large number. An authorized recipient has an inverse algorithm which also uses a key. An unauthorized receiver without the knowledge of the correct key has no chance to break the encipherment. More precisely, even if the encipherment algorithm is known, the probability that an unauthorized receiver may "guess" the key is very, very low with a good encipherment technique.

In ODA documents, the subordinates of a particular object – including associated content portions and maybe styles – can be enciphered as shown schematically in Fig. 41.

The enciphered document parts are removed from those parts of the document to which they originally belong (see Fig. 14) and moved as a single constituent into the protected document part. As a result, such subordinate constituents are no longer visible in the document. In particular, these constituents cannot be identified in the ODIF data stream.

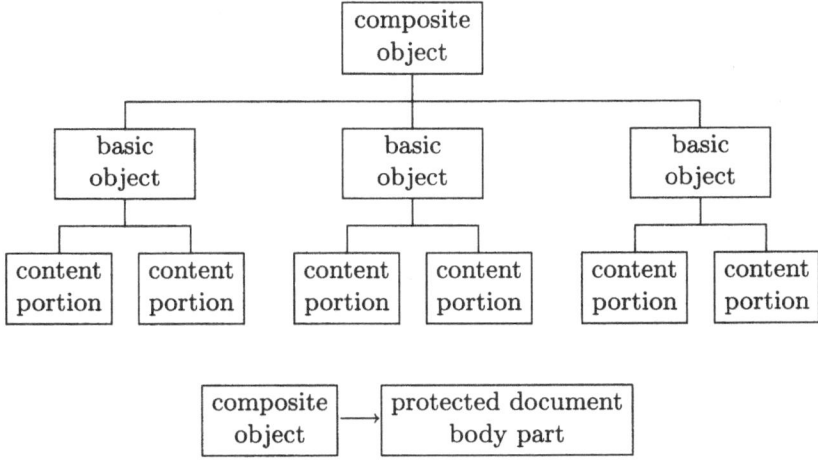

Fig. 41: Encipherment of parts of the document

In the example of Fig. 41 the attribute **enciphered** is added to the composite object. The parameter **protected part identifier** makes a reference to the constituent in the protected part which is created by the encryption process. The constituent in the protected part has only two attributes, one to specify its "name" for reference purposes and another one whose value is an octet string representing the enciphered constituents which have been removed from the unenciphered parts of the document (see Sect. 3.1.6). In other words, the complete structure below the composite objects is "compressed" into one single octet string. If the encipherment starts at the document logical root or document layout root level, the complete document can thus be protected from an unauthorized receiver. Even the document profile can be enciphered, at least the major part of it.

To permit an authorized receiver a decipherment of the encrypted parts, an entry is made in the document profile attributes **enciphered document profiles, pre-enciphered document body parts** or **post-enciphered document body parts** for each enciphered constituent in the protected part of a document. Such an entry contains sufficient information for an authorized receiver to start the decipherment algorithm. Of course, the correct key for decipherment is not included; this key is only known to an authorized receiver.

As the names "pre-enciphered document body part" and "post-enciphered document body part" for the constituents of the protected part indicate (see Fig. 15), there are two stages at which an encryption may be performed: either before the layout process starts or afterwards.

An authorized recipient receiving a document with a protected part, will firstly decipher all enciphered constituents for which he or she knows the keys. There may, however, remain some constituents for which a decipherment is not possible because access to them was only intended for another authorized receiver. In other words, the layout process and imaging process may have to deal with pre-enciphered and post-enciphered document body parts. Therefore, the Addendum on Security specifies the following rules to deal with such situations.

–   The layout process will ignore all pre-enciphered parts in a pre-enciphered processable document. In particular, the created layout structure will give no hint on the existence of such parts.
–   The imaging process will not display pre-enciphered and post-enciphered document parts, i.e., the content associated with them is not rendered on the presentation medium. However, since the positions and dimensions of the layout objects with subordinate post-enciphered objects are known, the areas of these objects will appear as "white space" on the presentation medium. In other words, the existence of pre-enciphered parts will not be visible in the imaged document, however, the existence of post-enciphered parts will be indicated by white areas.

Obviously, the editing process can only be applied to unenciphered parts of a document.

3.5.2   Seals

A so-called *seal* is a data item associated with a specified part of the document. The originator creates such seals in two steps:

–   Firstly, a so-called "fingerprint" (checksum) is generated by processing a particular part of a document using a specified algorithm. The main property of the algorithm is that it is computationally infeasible to construct another input to the algorithm resulting in the same output. In general, the fingerprint will be much shorter than the information it characterizes (i.e., in the order of bytes rather than kilobytes).
–   Secondly, the seal itself is created by taking the fingerprint – maybe with additional data such as the name or location of the originator – and applying a specified encryption algorithm to this data.

The authorized receiver may use the seals to verify the integrity and authenticity of the specified parts. For instance, the recipient may apply the same algorithm to create a checksum and test whether it is identical to the one provided by the originator. Of course, the precise method for the verification of a seal must only be known to an authorized receiver.

Seals may also be used for the purpose of non-repudiation of origin, i.e., by correctly decoding a seal a receiver may be able to prove to a third party that the seal must have been created by a particular originator.

An unauthorized receiver of a document is not able to verify integrity and authenticity and – even more important – he or she is not able to create correct seals. In particular, if an unauthorized person modifies the sealed parts of a document, an authorized receiver could detect such a modification since the seals would no longer be correct.

Readers familiar with UNIX[†] systems may know the program sum which is applied to a file. This program will usually – there exist several variants of the program – generate two numbers as "checksums". When electronically transmitting a file, the checksums created by sum are often also transmitted so a recipient may verify the integrity of the received file. (Of course, this is only possible if originator and recipient use the same variant of the program.)

For example, two persons might agree to always include seals in their interchanged documents by running the sum program on the specified parts of a document, multiply the two numbers generated by sum and add the number of beers they had at their last meeting in the pub. By inspecting the seals in a document, an unauthorized person would hardly detect the algorithm by which the seals were created, in particular, if he or she does not know the relation of the seals to beers and the sum program.

All seals associated with a document as a whole or with particular parts of it are specified in the document profile by means of the attributes ODA security label, sealed document profiles, pre-sealed document body parts and post-sealed document body parts. As the names of the attributes indicate, seals can be created before and after the layout process has been applied to a document.

Seals have no importance for the editing, layout or imaging process. They are only required to allow the security checks mentioned above by an authorized receiver. As soon as a sealed part of a document is modified, the associated seals become invalid.

[†] UNIX is a trademark of AT&T Bell Laboratories.

3.6   The Processing of ODA Documents

The processing of ODA documents is not described in detail in the ODA Standard. In particular, the Standard does not specify how an ODA document processing system should be implemented. Any system which can generate and receive a valid ODIF data stream can, in principle, be considered to conform to the Standard.

The Standard makes only a few comments on the editing process and briefly describes two "reference models" for the layout process and imaging process. It should be noted that the distinction between these three processes is mainly of a conceptual nature. In practice, this distinction will usually not be present, at least not from a user's point of view, since a native ODA system will certainly be implemented on a WYSIWYG ("what you see is what you get") basis, i.e., during the editing of a document the user will always see the document in formatted form: the editing, layout and imaging process are carried out instantaneously.

3.6.1   The Editing Process

The editing process is concerned with the creation and modification of ODA documents. According to the different constituents of a document, editing operations can be applied to

- the actual pieces of content represented by the values of the attribute content information for the individual content portions,
- the objects of the specific logical structure,
- the object classes of the generic structures,
- the layout styles,
- the presentation styles and
- the document profile.

Of course, from a user's point of view it is unacceptable that he or she has to edit the data structures of a document, i.e., the constituents and their attributes, directly. A direct creation and modification of these data structures is in practice impossible because of their complexity. Each ODA editor is required to hide these data structures from the user and provide a much more convenient user interface.

The editing of the specific logical structure and of the pieces of content is rather similar to the usual understanding of the term "editing a document". When editing a letter, for instance, an author creates a logical element "address" and creates or modifies the information attached to this logical object such as the postal address of a particular person.

For a full ODA implementation there are four different kinds of editors which can be distinguished for this kind of editing, at least conceptually: a *structure editor* for the creation, modification and deletion of the constituents of the specific logical structure and three *content editors* for the editing of character text, raster graphics and geometric graphics, respectively. If a generic logical structure belongs to a particular document, the structure editor must ensure that the specific logical structure is consistent with the generic logical structure (see also Sect. 3.2.15), i.e., a generic logical structure, if present, has to be taken into account during the editing process.

The editing of the generic logical structure is essentially the creation or modification of a document class description, in particular of the rules which apply to the logical structure of the documents belonging to the class. This concept is not very common among existing document processing systems and providing a convenient user interface for this kind of editing in an ODA system is not a trivial task.

The editing of the generic layout structure, the layout styles and presentation styles is essentially concerned with the intended representation of the document on a medium such as paper or a computer screen, since these constituents of an ODA document describe primarily the rules which are to be applied during the formatting and imaging of an associated document. In this case, the meaning of the term "editing" is quite different from its usual understanding and, again, the development of a convenient user interface may not be easy.

The information in the document profile will partially be created automatically by an ODA implementation, for instance, the number of pages or the content architectures used in the document. Certain kinds of information, however, cannot be generated automatically and have to be inserted manually by an author, for example, a summary of the intellectual content of a document or copyright specifications. The editing of the document profile will therefore rather likely be restricted to those pieces of information which cannot be derived automatically.

3.6.2 The Layout Process

The layout process creates the specific layout structure of a document – page sets, pages, frames and blocks, including the content portions associated with blocks – and may also create or modify some attributes of the document profile, for instance, the the attribute **number of pages** (see p. 193).

The ODA Standard describes only a so-called *reference model* of the layout process, i.e., a particular ODA system need not necessarily implement this model precisely. It is only required that the *result* of the layout process, i.e., the specific layout structure, conforms to the specifications in the logical structure, the generic layout structure, the layout styles and the presentation styles. The main concepts of this reference model shall be described in this section. It should be noted that the description will not include the case of basic pages. The concept of basic pages was primarily included in the ODA Standard for compatibility with existing CCITT Recommendations. Documents with basic pages are a rather special case of ODA documents and do not comprise the more powerful features of the Standard.

A closer look at the description of the layout process in the Standard shows that the specifications are at several places rather vague and may lead to different interpretations. From the present point of view it is rather likely that additional specifications on the layout process will be added in the future, especially when more practical experiences with the implementation of the Standard have been gained.

When creating the specific layout structure, the layout process takes the following constituents of a document into account:

- objects of the specific logical structure,
- object classes of the generic logical and generic layout structure,
- content portions associated with objects or object classes of the logical structure,
- content portions associated with object classes of the layout structure,
- layout styles and
- presentation styles.

It has to be distinguished between the *document layout process* which creates the objects of the specific layout structure, and the *content layout process* which performs the formatting of the content of the different content portions depending on their content architecture. (For simplicity, the term "formatting" is not only used for character content but also for raster graphics and geometric graphics content.) Part 2 of ISO 8613 describes only the document layout process; the different content layout processes are described in Parts 6, 7 and 8 which define the different content architectures. Unless explicitly stated otherwise, the term "layout process" in this section refers always to the document layout process, the different content layout processes are described in Sects. 6.4, 7.3 and 8.3.

Of course, the document layout process and the content layout processes interact with each other. From the view of the document layout process, the content of an ODA document consists only of a set of rectangular areas – the areas associated with frames and blocks – which have to be arranged on the individual pages of a document according to the specifications applying to the different objects. In particular, the internal structure of blocks – the layout objects at the lowest level containing character, raster graphics or geometric graphics content – are of no interest for the document layout process.

Slightly simplifying, the interaction between the document layout process and the content layout process can be viewed as follows:

- The document layout process determines rectangular areas into which the different pieces of content of a document can be laid out and informs a particular content layout process about the size of such an area. The content layout process then tries to format a piece of content into this area. If the content layout process does not succeed, for instance, if a raster image does not fit into the available area, it informs the document layout process about its failure. In this case, the document layout process would try to find a larger area and give the control back to the content layout process.
- If the content layout process could be carried out successfully, it informs the document layout process about the size of the area that is really needed for laying out the content. If this area is smaller than the area provided by the document layout process, the document layout process may decide how the unneeded area shall be used.

The construction rules for the generation of page sets, pages, frames and blocks are usually specified in the generic layout structure. All content of a document is formatted into blocks which can be created by the layout process in one of two ways:

- Firstly, blocks may be created by laying out the content associated with basic logical objects *without* the use of a layout object class of type block, i.e., a block is derived from the logical structure.
- Secondly, blocks may be created from a layout object class of type block, if this block class specifies content either by referencing a generic content portion or by means of the attribute **content generator**, i.e., a block is derived from the generic layout structure.

If the document contains a protected part (see Sect. 3.5), the pre-enciphered parts have to be deciphered before the layout process starts. The pre-enciphered parts which a receiver – even an authorized one –

cannot decipher are ignored. In particular, the created layout structure will give no hint on the existence of such parts.

The following two sections describe the methods for deriving the content associated with blocks, depending on the two different ways that blocks can be created.

Determining the content of basic logical objects

To determine the content associated with basic logical objects the following steps a) to m) are performed. As soon as one of these steps is successful, i.e., as soon as a piece of content has been determined, the remaining steps are ignored.

a) The basic logical object specifies the attribute content portions and for at least one of these content portions the attribute content information is specified. In this case, the content of the basic logical object is the value of the attribute content information. If the attribute content information is specified for more than one of the content portions referenced by the attribute content portions, the different pieces of content are concatenated according to their sequential order (see p. 91). The attributes which apply to each content portion, for instance, the attributes type of coding or alternative representation, are those specified for each individual content portion.

b) The basic logical object specifies the attributes content portions and content generator but none of the referenced content portions contains the attribute content information. The content associated with the basic logical object is then determined by evaluating the attribute content generator. The content portion attributes which apply are those specified by the first content portion in the sequential order.

c) The basic logical object specifies the attribute content portions but none of these content portions contains the attribute content information. The attribute content generator is not specified. In this case, the content associated with the basic logical object is an empty string; content portion attributes are not needed.

d) The basic logical object specifies the attributes content generator but not the attribute content portions. The content associated with the basic logical object is then determined by evaluating the attribute content generator. The value of the content portion attributes, for example, of type of coding or alternative representation are then determined according to the default mechanism described in Sect. 3.2.16.

e) The basic logical object references an object class by means of the attribute **object class**. This object class specifies the attribute **content portions** and for at least one of these content portions the attribute **content information** is specified. In this case, the content of the basic logical object is the value of the attribute **content information**. If the attribute **content information** is specified for more than one of the content portions referenced by the attribute **content portions**, the different pieces of content are concatenated according to their sequential order. The attributes which apply to each content portion are those specified for each individual content portion.

f) The basic logical object references an object class by means of the attribute **object class**. This object class specifies the attributes **content portions** and **content generator** but none of the referenced content portions contains the attribute **content information**. The content associated with the basic logical object is then determined by evaluating the attribute **content generator**. The content portion attributes which apply are those specified by the first content portion in the sequential order.

g) The basic logical object references an object class by means of the attribute **object class**. This object class specifies the attribute **content portions** but none of these content portions contains the attribute **content information**. The attribute **content generator** is not specified. In this case, the content associated with the basic logical object is an empty string; content portion attributes are not needed.

h) The basic logical object references an object class by means of the attribute **object class**. This object class specifies the attribute **content generator** but not the attribute **content portions**. The content associated with the basic logical object is then determined by evaluating the attribute **content generator**. The value of the content portion attributes are then determined according to the default mechanism described in Sect. 3.2.16.

i) The basic logical object references an object class by means of the attribute **object class** which in turn references an object class in a resource document by means of the attribute **resource**. This object class specifies the attribute **content portions** and for at least one of these content portions the attribute **content information** is specified. In this case, the content of the basic logical object is the value of the attribute **content information**. If the attribute **content information** is specified for more than one of the content portions referenced by the attribute **content portions**, the different pieces of content are concatenated according to their sequential order. The attributes which apply to each content portion are those specified for each individual content portion.

j) The basic logical object references an object class by means of the attribute **object class** which in turn references an object class in a resource document by means of the attribute **resource**. This object class specifies the attributes **content portions** and **content generator** but none of the referenced content portions contains the attribute **content information**. The content associated with the basic logical object is then determined by evaluating the attribute **content generator**. The content portion attributes which apply are those specified by the first content portion in the sequential order.

k) The basic logical object references an object class by means of the attribute **object class** which in turn references an object class in a resource document by means of the attribute **resource**. This object class specifies the attribute **content portions** but none of these content portions contains the attribute **content information**. The attribute **content generator** is not specified. In this case, the content associated with the basic logical object is an empty string; content portion attributes are not needed.

l) The basic logical object references an object class by means of the attribute **object class** which in turn references an object class in a resource document by means of the attribute **resource**. This object class specifies the attribute **content generator** but not the attribute **content portions**. The content associated with the basic logical object is then determined by evaluating the attribute **content generator**. The value of the content portion attributes are then determined according to the default mechanism described in Sect. 3.2.16.

m) If no content can be determined by one of the steps a)–l) the content associated with the basic logical object is an empty string; content portion attributes are not needed.

When the content of a basic logical object has been determined by one of these steps, it can then be formatted and a corresponding block can be created. It should be noted that at this stage all attributes which apply to the basic logical object and are relevant for the layout process, have to be taken into account. In particular, all attribute values that are expressions have to be evaluated.

Deriving the content of blocks from the generic layout structure

Besides the generation of blocks by content derived from the logical structure as described in the preceding section, blocks may also be created from the generic layout structure, for instance, if a frame class requires the creation of such a block by means of the attribute **generator for subordinates**. To analyze whether this is the case, the following nine steps are performed. As soon as the content of a block has been determined by one of these steps, the remaining steps are ignored.

a)   The block class which initiated the creation of the block contains the attribute **content portions** and for at least one of the referenced content portions the attribute **content information** is specified. The content associated with the block is then defined by the value of this attribute. If more than one content portion contains the attribute **content information**, this content is concatenated according to the sequential order of the content portions (see p. 96). The attributes which apply to the content portions, for example, the attributes **type of coding** or **alternative representation**, are those specified for each particular content portion.

   A content layout process, however, need not be performed for these content portions, since their content is already in formatted or formatted processable form, because the content portions belong to the generic layout structure.

b)   The block class which initiated the creation of the block contains the attribute **content portions** but none of the referenced content portions specifies the attribute **content information**. The attribute **content generator** is specified for the block class. In this case, the content associated with the block is derived by the evaluation of this attribute. The content portion attributes which apply are taken from the first content portion in the sequential order. In this case, a content layout process for the derived content has to be carried out and, in particular, a content portion is created.

c)   The block class which initiated the creation of the block contains the attribute **content portions** but none of the referenced content portions specifies the attribute **content information**. The attribute **content generator** is not specified for the block class. In this case, the content associated with the block is the empty string; content portion attributes are not needed.

d)   The block class which initiated the creation of the block references no content portions but the attribute **content generator** is specified for the block class. In this case, the content associated with the block is derived by the evaluation of this attribute. The content portion attributes which apply are derived according to the default mechanism described in Sect. 3.2.16. Again, a content layout process for the derived content has to be carried out.

e) The block class which initiated the creation of the block contains neither the attribute content portions nor the attribute content generator but refers to a block class in a resource document by means of the attribute resource. This block class contains the attribute content portions and for at least one of the referenced content portions the attribute content information is specified. The content associated with the block is then defined by the value of this attribute. If more than one content portion contains the attribute content information, this content is concatenated according to the sequential order of the content portions. The content portion attributes which apply are those specified for each particular content portion.

f) The block class which initiated the creation of the block contains neither the attribute content portions nor the attribute content generator but refers to a block class in a resource document by means of the attribute resource. This block class contains the attribute content portions but none of the referenced content portions specifies the attribute content information. The attribute content generator is specified for the block class. In this case, the content associated with the block is derived by the evaluation of this attribute. The content portion attributes which apply are taken from the first content portion in the sequential order.

g) The block class which initiated the creation of the block contains neither the attribute content portions nor the attribute content generator but refers to a block class in a resource document by means of the attribute resource. This block class contains the attribute content portions but none of the referenced content portions specifies the attribute content information. The attribute content generator is not specified for the block class. In this case, the content associated with the block is the empty string; content portion attributes are not needed.

h) The block class which initiated the creation of the block contains neither the attribute content portions nor the attribute content generator but refers to a block class in a resource document by means of the attribute resource. This block class references no content portions but the attribute content generator is specified for the block class. In this case, the content associated with the block is derived by the evaluation of this attribute. The content portion attributes which apply are derived according to the default mechanism described in Sect. 3.2.16.

i)   If none of the above steps a)–h) succeed, the content associated with the block is the empty string; content portion attributes are not needed.

It should be noted that these nine steps are not only performed for blocks but also for basic pages. However, as mentioned above, documents with basic pages shall not be considered here.

Determining the content of frames

The preceding two sections defined the rules according to which the content associated with blocks is determined. Blocks are usually assembled into frames and, as a consequence, the content of frames is determined by the content of their subordinate blocks. However, there exists an additional way to create content of a frame, namely by means of the attribute **logical source**. An example is given with the description of this attribute on p. 106.

Therefore, whenever a new frame is created during the layout process, it is analyzed whether one of the following two cases applies:

a)   The frame refers by means of the attribute **object class** to a frame class for which the attribute **logical source** is specified. In this case, all logical objects defined by the value of this attribute are created and the content associated with the generated basic logical objects is determined according to the rules given above.

b)   The frame refers to a frame class which in turn refers to a frame class in a resource document by means of the attribute **resource**. The frame class in the resource document specifies the attribute **logical source**. In this case, all logical objects defined by the value of this attribute are created and the content associated with the generated basic logical objects is determined according to the rules given above.

If neither a) nor b) holds, the content associated with the frame is the content associated with the blocks subordinate to the frame.

Available area

The rectangular area which the document layout process supplies to a content layout process to lay out the content of a block is called the *available area*.

The first restriction on the size of this area is given by the size of the area of the frame which is immediately superior to the block: the area of a block must be completely contained within the area of its superior frame. If the frame already contains blocks, the available area for a new block is further diminished by the areas consumed by these blocks.

The remaining area may be further reduced, in particular by the area needed for

– a border, if the attribute border is specified for the frame or the block (see Fig. 32);
– the offsets to the edges of the frame, if the attribute offset is specified (see Fig. 29);
– the offsets to adjacent blocks, if the attribute separation is specified (see Fig. 30). In this case, the value of the attribute fill order (see p. 112) has also to be taken into account.

If the attribute concatenation is specified for a basic logical object, i.e., if the content associated with the object shall be displayed in an already previously formatted block, a new available area is determined for this block and the content layout process is carried out again, taking into account, of course, the additional content which is concatenated to the preceding content portion.

Size and position of frames and blocks

As one of the main results of the layout process the horizontal and vertical sizes of the specific layout objects, i.e., of pages, frames and blocks, and their positions within their superior layout objects (pages or frames) are determined. In particular, the values of the two sub-parameters fixed position of the attribute dimensions (see p. 119) and the values of the sub-parameters horizontal position and vertical position of the attribute position (see p. 117) have to be determined.

This is easy for pages: The size of a page is either specified in the generic layout structure or a default value is derived according to the procedure described in Sect. 3.2.16. It is also easy for frames or blocks which are immediately subordinate to a page: Their sizes are also either specified in the generic layout structure or their default value is the size of the page. Similarly, the position of such frames or blocks is specified in the generic layout structure or the default value for the attribute position is used.

The situation is more complicated for frames or blocks which are *not* immediately subordinate to a page. Their size or position may be specified explicitly in the generic layout structure, but this need not necessarily be the case. Besides the formatting of the actual content of a document, the typical task of the layout process is, in fact, the determination of column breaks and page breaks and the positioning of the layout objects on the pages of a document.

If the size of a block is not explicitly specified, it is determined by the content layout process depending on the size required for holding the formatted piece of content associated with the block. The position of a block within its superior frame is either explicitly defined – if the parameter fixed position of the attribute position is specified – or it is derived by the value of the attribute layout path, taking into account the attributes border, offset, separation and fill order.

If the size of a frame is not explicitly specified, it is defined by the sizes required by the subordinate layout objects (frames or blocks). It should be noted, however, that for the size of a frame only a preliminary value may be available during the layout process, for instance, when at a certain stage during the layout process several blocks have been already assigned to a frame but additional blocks will be added later on when they are created. The position of a frame within its superior frame is either explicitly defined – if the parameter fixed position of the attribute position is specified – or it is defined by the sub-parameters offset, separation, alignment and fill order of the attribute position. Furthermore, the attribute layout path and border have to be taken into account.

Interactions and precedences among layout directives

The attributes block alignment, concatenation, fill order, indivisibility, layout category, layout object class, new layout object, offset, same layout object, separation and synchronization which can be specified for layout styles are called *layout directives* in the ODA Standard, since they specify rules on how the layout process should be carried out for the logical objects to which they apply. Though the specification of a particular layout directive may be sensible in general for a certain set of objects, there may arise situations where a layout directive cannot be satisfied.

Consider, for example, the case that all content of individual subsections of a document shall be printed on a single page. These subsections shall be small enough that this requirement can usually be satisfied, i.e., each of these subsections fits usually on one page. This could be achieved in an ODA document by specifying the attribute indivisibility (see p. 130) for a layout style which is referenced by the respective logical objects. If, however, a certain subsection contains content which would not fit on a single page, this layout directive cannot be satisfied.

Such kinds of conflicts – that a generally desirable layout cannot be achieved in a particular situation – is a general problem of document processing and not specific for ODA documents. Therefore, the ODA Standard specifies a set of precedence rules and interactions among layout directives. If not all of the layout directives can be satisfied, the layout process shall try to satisfy the directives according to their importance which is – in decreasing order – defined as follows:

1. **layout object class**: this attribute need not be considered, if its value is null;
2. **layout category** (only applicable to basic logical objects): this attribute need not be considered, if its value is null;
3. **new layout object**: is ignored, if the attribute **layout object class** applies and specifies a so-called *lowest level frame* (see p. 53); need not be considered, if its value is null;
4. **same layout object**: is ignored, if the attribute **layout object class** applies and specifies a lowest level frame; need not be considered, if its value is null;
5. **fill order** (only applicable to basic logical objects);
6. **concatenation** (only applicable to basic logical objects);
7. **offset** (only applicable to basic logical objects);
8. **separation** (only applicable to basic logical objects);
9. **synchronization**: this attribute need not be considered, if its value is null;
10. **indivisibility**: is ignored, if the attribute **layout object class** applies and specifies a lowest level frame;
11. **block alignment** (only applicable to basic logical objects).

It should be noted that these layout directives shall not only be satisfied locally, for instance, when processing all basic logical objects subordinate to a particular composite logical object. A layout which satisfies the layout directives at a certain hierarchical level in the specific layout structure must not violate layout directives specified at higher hierarchical levels. In case of conflict, layout directives at a higher level take precedence over those at lower levels.

Layout streams

The ODA Standard contains the concept of the so-called *layout streams*. The rationale for this concept is as follows: The layout process progresses the logical objects and their associated content portions in their sequential order (see p. 91) and creates the corresponding specific layout objects accordingly. In several cases, however, the sequential order of the layout objects should be different from the order in which the logical objects are processed.

For example, an ODA document may consist of several chapters, and a table of contents – listing the chapter headings and maybe page numbers – is to be created automatically. The table of contents is to be printed on the first page of a document. Each time the layout process starts the processing of a new chapter, an additional entry therefore has to be made on the first page of the document. In other words, during the processing of the document, the objects which are created in the specific layout structure are directed to rather distinct places, i.e., the content flows into different "layout streams". (An example is given after the description of the attribute **permitted categories** on p. 128).

The initialization of different layout streams is achieved by assigning a so-called *layout category* to basic logical objects by specifying the attribute **layout category** (see p. 128) for a layout style which is referenced by the basic logical object. The value of the attribute **layout category** is a character string from the minimum subrepertoire of ISO 6937, Part 2, which can be regarded as the "name" of the layout stream assigned to the basic logical object. Each different name which appears in an ODA document generates a separate layout stream. For instance, an ODA document might contain layout streams for a table of contents, a list of figures and an index.

Whenever a basic logical object is assigned to a layout stream, its content is not displayed in the specific layout structure at that position which reflects its sequential order in the logical structure, but the layout process determines the position which is reached for the particular layout stream in the specific layout structure.

Concerning the layout structure, layout streams can only be assigned to frames whose immediately subordinate objects are blocks since only basic logical objects can have an associated layout stream. The assignment of layout streams to frames, i.e., the specification, into which kind of frame the content of a particular layout stream shall be "poured", is done by specifying the attribute **permitted categories** (see p. 128) for a layout style which is referenced by the frame. This attribute specifies one or more names of layout streams whose content may be displayed in the frame.

If one or more layout streams are initialized in a document, each layout stream has a so-called *current layout position*. The content of a basic logical object which has an associated layout stream is always inserted at the current layout position of the respective layout stream. In particular, the sequential order of the layout objects belonging to a specific layout stream reflect always the sequential order of the basic logical objects of this layout stream.

3.6.3   The Imaging Process

The imaging process is the final processing step for ODA documents described in the Standard. It uses the specific layout structure with associated content portions, the generic layout structure, if it contains generic content portions, and presentation styles, if present, as its input and creates a humanly perceivable representation of the document, usually on paper or on a computer screen.

If the document contains a protected part (see Sect. 3.5), the post-enciphered document body parts have to be deciphered before the imaging process starts if the receiver is able to perform this decipherment. The imaging process will not display pre-enciphered and post-enciphered document parts, i.e., the content associated with them is not rendered on the presentation medium. However, since the positions and dimensions of the layout objects with subordinate post-enciphered objects are known, the areas of these objects will appear as "white space" on the presentation medium. In other words, the existence of pre-enciphered parts will not be visible in the imaged document, however, the existence of post-enciphered parts will be indicated by white areas.

Since the imaging process is obviously very dependent on the rendition hardware, the ODA Standard contains only rather few specifications – partially in Part 2, partially in those Parts which define the different content architectures (see Sects. 6.5, 7.4 and 8.4) – on how the imaging process should be performed. This section describes the specifications of Part 2 which are generally valid, independently of a particular content architecture.

The order in which the pieces of content of a document are rendered on a presentation medium is either defined by the attribute **imaging order** which can be specified for composite pages or frames, or – if this attribute is missing – by the sequential order of the layout objects (see p. 91). This imaging order is important since different orders may lead to different results depending on the attribute **transparency** (see p. 133).

The areas of frames, blocks and pages have an associated texture which is defined by the attributes colour and transparency. The following combinations are possible:

- colour = white, transparency = opaque,
- colour = colourless, transparency = opaque and
- colour = colourless, transparency = transparent.

(The combination colour = white, transparency = transparent is not permitted.)

If the areas of two layout objects intersect, the content and the texture of the upper area hides the content and texture of the lower area, if for the upper layout object the attribute transparency has the value opaque. If the value is transparent, content and texture of both areas overlay and, in particular, the content of both areas is visible.

Concerning the imaging process, it must also be distinguished whether the rendition medium is a sheet of paper or a computer screen. For a computer screen, the parameters side of sheet and nominal page size can be ignored since such a screen has no "recto" and "verso" side and the concept of a "nominal page size" is also not very sensible. It is implementation dependent how these parameters are handled. For instance, if a complete ISO A4 page cannot be imaged on a screen, an implementation may provide facilities to scroll the page on the screen.

Determining the content of basic layout objects during the imaging process

During the imaging process, i.e., when the final rendition of a document on a presentation medium is created, the content associated with the basic layout objects is determined according to the following four steps. As soon as a piece of content is derived by one of these steps, the remaining steps are ignored.

a)   One or more content portions are associated with the basic layout object by means of the attribute content portions. In this case, the content which is displayed in the area of the basic layout object is determined by evaluation of the attribute content information for each individual content portion. If more than one content portion contains this attribute, the pieces of content are concatenated.

b)   The basic layout object refers to a layout object class which has one or more associated content portions. In this case, the content which is displayed in the area of the basic layout object is determined by evaluation of the attribute **content information** for these content portions. Again, the content of more than one content portion is concatenated.

c) The basic layout object refers to a layout object class which refers to a layout object class in a resource document. This object class has one or more associated content portions. In this case, the content which is displayed in the area of the basic layout object is determined by evaluation of the attribute **content information** for these content portions and, if necessary, the content of more than one content portion is concatenated.

d) If none of the three preceding cases applies, the content associated with the basic layout object is an empty string, i.e., no content appears in the area of the basic layout object.

In the first three cases, the content portion attributes of the individual content portions apply; no content portion attribute are needed in the last case.

Rendition on paper

Concerning the rendition of ODA documents on paper, the Standard contains a set of specifications which address the problem of the different paper sizes which are currently used in different areas of the world. Besides the ISO A4 size and – to a lesser extent – A3 size which are common in Europe, the *North American Letter Size, Japanese Letter Size* and *Japanese Legal Size* are frequently used in other areas. These are also the sizes which today's printer equipment such as laser printers can normally process.

In practice, however, a printer is usually not able to print on each area of a sheet of paper. Printing is almost always impossible in certain areas along the edges of the paper, for instance, because of gripping losses for paper feeding. The ODA Standard therefore introduces the concept of a so-called *assured reproduction area* which denotes that area on a sheet of paper which remains on the nominal page after deducting allowances for edges losses, as shown in Fig. 42.

Firstly, it is expected that each printer used for the rendition of ODA documents can at least print within the assured reproduction area. Secondly, the author of an ODA document should make sure that all content of his document falls within this area. The actual values for the sizes of the assured reproduction areas and their distances to the edges of the paper depending on the paper size are given in Fig. 43.

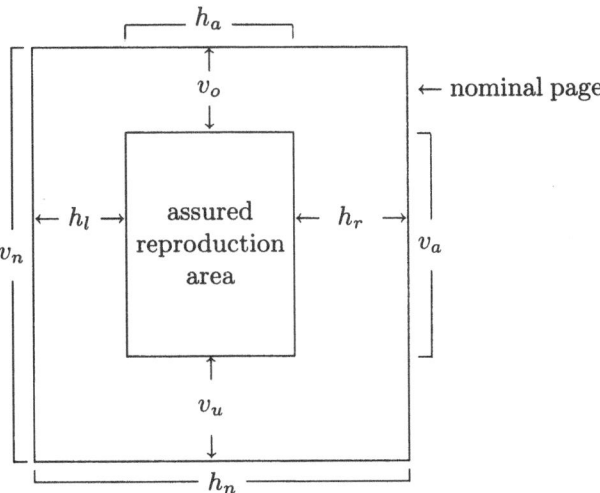

Fig. 42: Position of the assured reproduction area on a page

	ISO A4	ISO A3	North Amer-ican Letter	Japapese Letter	Japanese Legal
h_n	9920 BMU	14030 BMU	10200 BMU	8598 BMU	12141 BMU
	210.0 mm	297.0 mm	215.9 mm	182.0 mm	257.0 mm
v_n	14030 BMU	19840 BMU	13200 BMU	12141 BMU	17196 BMU
	297.0 mm	420.0 mm	279.4 mm	257.0 mm	364.0 mm
h_a	9240 BMU	13200 BMU	9240 BMU	7600 BMU	11200 BMU
	195.6 mm	279.4 mm	195.6 mm	160.9 mm	237.1 mm
v_a	13200 BMU	18480 BMU	12400 BMU	10200 BMU	15300 BMU
	279.4 mm	391.2 mm	262.5 mm	215.9 mm	323.8 mm
h_l	345 BMU	345 BMU	345 BMU	400 BMU	400 BMU
	7.3 mm	7.3 mm	7.3 mm	8.5 mm	8.5 mm
h_r	335 BMU	485 BMU	615 BMU	598 BMU	541 BMU
	7.1 mm	10.3 mm	13.0 mm	12.7 mm	11.5 mm
v_o	472 BMU	472 BMU	472 BMU	900 BMU	900 BMU
	10.0 mm	10.0 mm	10.0 mm	19.0 mm	19.0 mm
v_u	358 BMU	888 BMU	328 BMU	1041 BMU	996 BMU
	7.6 mm	18.8 mm	6.9 mm	22.0 mm	21.1 mm

Fig. 43: Size and position of the assured reproduction area for different paper sizes

The (physical) size of a sheet of paper of an ODA document is defined by the parameter **nominal page size** of the attribute **medium type**. The

(logical) size of a page in an ODA document is defined by the attribute
dimensions. It is therefore recommended that the value of the attribute
dimensions does not exceed the size of the assured reproduction area of a
particular nominal page. (Of course, an ODA document need not necessarily specify one of the five standard pages sizes shown in Fig. 43, but
the usage of other page sizes might lead to additional difficulties for a
receiver.)

The position of a (logical) page on a sheet of paper is defined by the
attribute **page position** (see p. 134). If this attribute is not explicitly
specified, the page is positioned according to the following rules:

- If the size of the page does not exceed the size of the assured reproduction area, the upper left corner of the page is placed at the position
  given by the coordinates (h_l, v_o), i.e., at the upper left corner of the
  assured reproduction area.
- If the size of the page is the same as the size of the sheet of paper,
  i.e., if the value of attribute **dimensions** for the page is the same as
  the value of the parameter **nominal page size** of the attribute **medium
  type**, the upper left corner of the page is placed at the position $(0,0)$.
- Otherwise, the ODA Standard specifies that the value of the attribute **page position** shall be chosen in such a way that "the pos-
  sibility of information loss in minimized". This is considered to be
  implementation-dependent.

It should be noted that the ODA Standard supports not only printing
in "portrait mode" (the vertical dimension is greater than the horizontal
one) but also in "landscape mode" (the horizontal dimension is greater
than the vertical one). Of course, the usage of landscape mode requires
that a printer supports this feature.

4 Part 4: The Document Profile

This Part of the ODA Standard contains the description of the document profile which is always a constituent of an ODA document.

The document profile contains information about the document to which it belongs. A part of this information is required for the processing of the document, for instance, the identifier of a resource document (see Sect. 3.4.3) which may be referenced by constituents of the document by means of the attribute resource.

Other information in the document profile may not directly be necessary to process the document but it may be an indication which features an implementation has to support to be able to handle it. Such information may be particularly important in the context of document application profiles, i.e., for an ODA system which has only implemented a subset of the ODA Standard. For instance, the document profile specifies which content architectures are used in the document.

Some information in the document profile is of no practical value for the automatic processing of the document, for instance, the name of the author of the document. However, such information may be of great importance for the "user" of the document.

When interchanging an ODA document, the information in the document profile is encoded within the ODIF data stream like all the other information in the document. The representation of the document profile follows the same principles that hold for the document body. In particular, a document profile is a separate constituent in the document consisting of a set of attributes.

It should be pointed out once again that the document profile alone without the document body which belongs to it, can be electronically interchanged (see Sect. 3.3.2). A receiving system may decide from the information in the document profile, whether it can process the document and whether therefore an interchange of the complete document is useful.

4.1   Constituents of the Document Profile

As mentioned above, a document profile is a separate constituent of a document consisting of a set of attributes. The mandatory, defaultable and non-mandatory attributes are as follows:

document profile :=
  {content architecture classes, document architecture class,
   document reference, interchange format class, ODA version,
   ⟨unit scaling⟩,
   [abstract], [access rights], [additional information],
   [alternative feature sets], [alternative representation character sets],
   [authorization], [authors], [block alignments],
   [borders], [coding attributes], [colours],
   [comments character sets], [copyright], [creation date and time],
   [distribution list], [document application profile],
   [document application profile defaults], [document date and time],
   [document size], [document type], [enciphered document profiles],
   [enciphered profiles], [expiry date and time],
   [external document class], [fill orders], [fonts list],
   [generic layout structure], [generic logical structure],
   [keywords], [languages], [layout paths], [layout styles],
   [local file references], [local filing date and time],
   [medium types], [number of objects per page],
   [number of pages], [ODA security label], [organizations],
   [owners], [page dimensions], [page positions],
   [post-enciphered body parts], [post-enciphered document body parts],
   [post-sealed document body parts], [pre-enciphered body parts],
   [pre-enciphered document body parts],
   [pre-sealed document body parts], [preparers], [presentation features],
   [presentation styles], [profile character sets], [protections],
   [purge date and time], [references to other documents],
   [release date and time], [resource document], [resources],
   [revision history], [sealed document profiles], [sealed profiles],
   [security classification], [specific layout structure],
   [specific logical structure], [start date and time], [status], [subject],
   [superseded documents], [title], [transparencies], [types of coding],
   [user-specific codes] }

Only six attributes are classified as mandatory or defaultable, namely content architecture classes, document architecture class, document reference, interchange format class, ODA version and unit scaling. The first five attributes must always be specified. If the last attribute is missing, a default value will be assumed.

There are, however, some further attributes whose presence depends on the properties of the document itself. These are the attributes generic layout structure, generic logical structure, specific layout structure, specific logical structure, layout styles, presentation styles, external document class, resource document, resources, enciphered profiles, sealed profiles, post-enciphered body parts and pre-enciphered body parts.

The attributes generic layout structure and generic logical structure have to be specified if and only if the document body contains a generic layout structure or generic logical structure, respectively. The values of the attributes specify whether the particular structure is a complete generator set, partial generator set or factor set (see Sect. 3.2.15).

Similarly, the attributes specific layout structure and specific logical structure have to be specified if and only if the document body contains a specific layout structure or a specific logical structure, respectively. Theses two attributes have then the value present. Furthermore, the attributes layout styles and presentation styles with the value present have to be specified, if the respective styles appear in the document body.

If the generic structures of a document are stored in an *external document class* (see Sect. 3.4.2), the attribute external document class must be specified in the document profile. Likewise, the attribute resource document must be present, if a *resource document* is referenced within the document body by means of the attribute resource (see Sect. 3.4.3).

Furthermore, the attributes enciphered profiles, sealed profiles, post-enciphered body parts and pre-enciphered body parts have to be specified if and only if the document body contains enciphered document profiles, sealed document profiles, post-enciphered document body parts or pre-enciphered document body parts, respectively. Theses attributes have then the value present.

It should be noted that the attribute alternative feature sets was introduced by the Addendum on Alternate Representations and the attributes enciphered document profiles, enciphered profiles, ODA security label, post-enciphered body parts, post-enciphered document body parts, post-sealed document body parts, pre-enciphered body parts, pre-enciphered document body parts, pre-sealed document body parts, sealed document profiles and sealed profiles by the Addendum on Security; they were not contained in the initially published version of the Standard.

4.2   The Attributes of the Document Profile

In this section the attributes of the document profile shall be described in detail, in particular their value ranges and their semantics. The description is structured into subsections according to the semantics of the attributes.

4.2.1   Data Types for the Attribute Values of the Document Profile

Besides the data types for attribute values which were introduced already in Sect. 3.2.2, there are some other data types for attributes of the document profile which are described in this section.

For several attributes of the document profile the value consists of a sequence of characters. The character repertoire for these character strings is defined by the attribute **profile character sets** (see p. 189). Such a character string shall be called a *profile character set string* and be considered a separate data type.

Another data type shall be introduced for the format of names to identify a particular person in certain attributes of the document profile. Such a name is called a *personal name* in the Standard, defined in Annex A of ISO 8613, Part 4. The specifications are taken from the CCITT Recommendation X.411.

A *personal name* has up to four parameters called **surname**, **givenname**, **initials** and **title**, where the value of each parameter is a *profile character set string*. The parameter **surname** specifies the family name of the person; this parameter is mandatory. The parameter **givenname** is the name by which the person is commonly known; this parameter is optional. The value of the parameter **initials** consists of a sequence of initial characters and the value of the parameter **title** is a title by which a person is normally addressed in documents; these two parameters are also optional. This can be summarized as follows:

$$
\begin{aligned}
\textit{personal name} := \{ &\textsf{surname} = \text{ '}\textit{profile character set string}\text{'} \\
&[\textsf{givenname} = \text{ '}\textit{profile character set string}\text{'}] \\
&[\textsf{initials} = \text{ '}\textit{profile character set string}\text{'}] \\
&[\textsf{title} = \text{ '}\textit{profile character set string}\text{'}]\}
\end{aligned}
$$

Several attributes of the document profile contain time and date specifications. The format of these specifications is as defined by the ISO

Standard 8601 (*Information interchange – Representation of dates and times*) and – since the format is defined in an external Standard – a separate data type called *date and time* shall be introduced for it. For attributes using this data type, it is always sufficient to specify the calendar date; an additional specification of the time of day is optional.

The attribute fonts list (see p. 205) references particular fonts (graphic character sets). The format of such references is defined in the ISO Standard 9541, Part 5 (*Font and character information interchange – Part 5: Font attributes and character model*). Since this format is also described in a different Standard, a separate data type called *font reference* shall be introduced.

Several attribute values relating to character sets specify escape sequences as defined in ISO Standard 2022. Therefore, the data type *ISO 2022 escape sequences* shall also be introduced as an additional elementary data type in ODA documents.

Some parameters of attributes in the document profile relating to security aspects contain data elements which are required for an authorized recipient to deduce a key which is needed to verify a seal or decipher enciphered data elements. Since sealing or encryption algorithms are not defined in the ODA Standard, the structure of the values of these parameters is application dependent. The data type of these values shall be called *key information*.

4.2.2   Specification of Implementation Requirements

The five attributes ODA version, interchange format class, content architecture classes, number of objects per page and alternative feature sets can be regarded as specifications of requirements which an ODA implementation has to satisfy for processing the document belonging to the document profile.

The attribute ODA version identifies the version of the ODA Standard to which the document conforms. At present, there is only one version, namely the ISO Standard 8613 as published in 1989 (and the compatible CCITT Recommendations of the T.410 series published in 1988) and therefore the attribute is not yet very useful. It will, however, become meaningful, when addenda or modifications of the Standard are published in the future which require modifications of ODA implementations.

The value of this attribute consists of a sequence of two entries. The first entry is a *profile character set string* with the name of the Standard (for instance, "ISO 8613"), the second entry is a calendar date according to ISO 8601 which indicates that the document conforms to the version and addenda published up to this date. This can be summarized as:

ODA version $=$ ⟦ '*profile character set string*' '*calendar date*' ⟧

The attribute ODA version must always be specified.

The attribute interchange format class specifies whether the ODIF data stream is encoded according to the class A or B (see Sect. 5.2), i.e.:

interchange format class $=$ (A | B)

The attribute interchange format class must always be specified.

The attribute content architecture classes is a list of all content architecture classes used in the document, i.e., each particular value for the attribute content architecture class (note the missing "es" at the tail) which appears somewhere in the document, is included in this list. According to Parts 6, 7 and 8 of the Standard, permissible values for the attribute content architecture class are the following ASN.1 object identifiers (see p. 239, 273 and 305): "2 8 2 6 0" (for formatted character content), "2 8 2 6 1" (for processable character content), "2 8 2 6 2" (for formatted processable character content), "2 8 2 7 0" (for formatted raster graphics content), "2 8 2 7 2" (for formatted processable raster graphics content) and "2 8 2 8 0" (for formatted processable geometric graphics content). We therefore have:

content architecture classes $=$
    {['2 8 2 6 0'] ['2 8 2 6 1'] ['2 8 2 6 2']
     ['2 8 2 7 0'] ['2 8 2 7 2'] ['2 8 2 8 0']}

The attribute content architecture classes must always be specified. Though all values in the list are optional, at least one must be present since obviously at least one content architecture must appear in the document.

The attribute number of objects per page indicates the maximum number of specific layout objects appearing per page in the document. Its value is a positive integer:

number of objects per page $=$ '*positive integer*'

This attribute is only specified if the number of objects per page exceeds the value specified by a document application profile to which the document conforms. The attribute can also be regarded as an implementation requirement: though the ODA Standard in general defines no upper bound on the number of layout objects on a page, each ODA implementation will probably only be able to handle a certain number of such objects, for instance, during the layout process.

The value of the attribute alternative feature sets – this attribute was included by the Addendum on Alternate Representations – is a set of sets where the elements of the subordinate sets are ASN.1 object identifiers:

$$\text{alternative feature sets} = \left\{ \left[\{ASN.1 \ object \ identifier\} \right]^{+} \right\}$$

Each of the subordinate sets contains the ASN.1 object identifiers of those features which are sufficient to process a particular set of alternative descriptions in the document. ASN.1 object identifiers have been chosen for the specification of features since many of them, for instance, the content architecture classes, are identified by such object identifiers.

For example, a particular document may contain formatted character content and formatted processable geometric graphics content as primary descriptions, and, furthermore, each basic object representing a geometric graphics picture may have a corresponding raster graphics image in formatted form as an alternative description. The value of the attribute alternative feature sets in the document profile could then be

$$\{\{\text{``2 8 2 6 0''} \ \text{``2 8 2 8 0''}\} \ \{\text{``2 8 2 6 0''} \ \text{``2 8 2 7 0''}\}\}$$

to indicate that a receiving system can process the document correctly, if it supports either the combination of formatted character content ("2 8 2 6 0") and formatted processable geometric graphics content ("2 8 2 8 0") or of formatted character content ("2 8 2 6 0") and formatted raster graphics content ("2 8 2 7 0").

4.2.3   Specification of the Document Structures

Seven attributes of the document profile specify which structures appear in the document body and to which document architecture class the document belongs. These are the attributes generic layout structure, generic logical structure, specific layout structure, specific logical structure, layout styles, presentation styles and document architecture class.

The attributes generic layout structure and generic logical structure must be specified if the corresponding structures appear in the document body. Their values indicate whether the structure is a complete generator set, partial generator set or factor set (see Sect. 3.2.15). We therefore have:

generic layout structure =
  (complete generator set | partial generator set | factor set)

generic logical structure =
  (complete generator set | partial generator set | factor set)

It should be noted that the values of these two attributes not only give useful information about the document in general, but have to be taken into account during the processing of the document. As explained in Sect. 3.2.15, consistency requirements between the generic structures and specific structures depend on whether a generic structure is a complete generator set or partial generator set, and – depending on the consistency requirements – a particular modification of the document may be permitted or prohibited.

In the same manner, the attributes specific layout structure, specific logical structure, layout styles, presentation styles, enciphered profiles, sealed profiles, post-enciphered body parts and pre-enciphered body parts have to be specified in the document profile, if the document body contains the respective specific structures, styles or constituents of the protected document part. The value of these attributes is always present:

specific layout structure = present

specific logical structure = present

layout styles = present

presentation styles = present

enciphered profiles = present

sealed profiles = present

post-enciphered body parts = present

pre-enciphered body parts = present

The attribute document architecture class indicates whether the document is in formatted, processable or formatted processable form (see Sect. 3.3.1):

document architecture class =
    (formatted | processable | formatted processable)

This attribute must always be specified and its value must, of course,
reflect the actual structure of the document. For instance, its value must
not be formatted, if the document body contains no specific layout structure which is always required for a formatted document.

4.2.4   References to Other Documents

As explained in Sect. 3.4, certain components of an ODA document
may be stored separately and are only used when the document is to
be processed. Such components – which have also a document profile
and are considered valid documents in their own right – are called *exter-*
nal document class descriptions and *resource documents* (see Sects. 3.4.2
and 3.4.3)

If a documents refers to an external document class description or to a
resource document, it must specify the attributes external document class
or resource document, respectively, in its profile. The value of these two
attributes is either a so-called *ASN.1 object identifier* (see p. 209) or a
profile character set string:

external document class =
    (*'ASN.1 object identifier'* | *'profile character set string'*)

resource document =
    (*'ASN.1 object identifier'* | *'profile character set string'*)

The value of the attributes is identical to the value of the attribute
document reference (see p. 198) in the document profile of the external
document class description or resource document which is referenced.

The generic structures or styles within a external document class are
only used during the processing of a particular document, if the document
does not contain these generic structures or styles. If they are present,
generic structures or styles in the external document class are ignored
(see Sect. 3.4.2). Therefore, no conflict can arise between constituents
in a document and corresponding constituents in a referenced external
document class.

The situation is different for resource documents, which are usually
applied to provide additional attribute values during the processing of
a document, especially for the default mechanism (see Sect. 3.2.16). In

particular, a resource document contains objects classes which are to be used *in addition* to the object classes already present in the referencing document and both object classes use *object-class-ids* as values for the attributes object class identifier.

The ODA Standard allows the use of the same *object-class-ids* both in a particular document and in a resource document which is referenced. This allows a greater degree of independence between a document and a resource document. (It should be noted that a resource document may be referenced by several documents.) For instance, a reference to a resource document may be made at a certain stage during the editing of the document, when *object-class-ids* have been assigned in the document which might already have been assigned in the resource document. To avoid conflicts, the reference to an object class in a resource document by means of the attribute resource is done by using a "symbolic name" consisting of a character string from the minimum subrepertoire of ISO 6937, Part 2 (see p. 98).

The mapping of these symbolic names is done in the document profile of a document referencing a resource document by means of the attribute resources. (Note the "s" at the tail of the document profile attribute.) The value of the attribute resources is a set of one or more elements where each element is an *ISO 6937/2 character string* – this string is used as the value for the attribute resource within the document – and an *object-class-id* – this is the *object-class-id* used in the resource document, i.e.:

resources =
$$\left\{\left[\!\left[\,'\textit{ISO 6937/2 character string}'\ \textit{object-class-id}\,\right]\!\right]^{+}\right\}$$

The following example may illustrate this concept. Within a document the attribute resource may be specified for a particular constituent in the form:

resource = 'abc'

A mapping of this symbolic name used for specifying the document class in the resource document is then performed by the attribute resources in the document profile, for instance, in the form:

resources = {'abc' [[1 3 4]]}

As a result, the object class in the resource document having the *object-class-id* [[1 3 4]] is referenced by the symbolic name "abc" from within the document. (Of course, the resource document must contain an object class with this *object-class-id.*)

If a document conforms to a particular document application profile, the attribute **document application profile** is specified in the document profile. Its value is either an ASN.1 object identifier or the integer 2:

document application profile = (*'ASN.1 object identifier'* | 2)

The value "2" was included for compatibility with existing CCITT Recommendations and denotes the document application profile for Group 4 Facsimile documents as defined in Recommendation T.503.

4.2.5   Specification of Character Sets

Three attributes of the document profile – **profile character sets**, **alternative representation character sets** and **comments character sets** – indicate character sets used for certain character strings which may be the values for some attributes of the document profile and of other constituents in the document body.

The attribute **profile character sets** defines the character set which is used for those attributes of the document profile whose value is a *profile character set string* (see p. 182). The attribute **alternative representation character sets** indicates the character set for the value of the attribute **alternative representation** which can be specified for content portions, and the attribute **comments character sets** defines the character set for values of the attributes **user-readable comments** and **user-visible name**.

The values of the three attributes consist of the escape sequences used to announce and designate the character sets according to ISO 2022, i.e.:

profile character sets = *'ISO 2022 escape sequences'*

alternative representation character sets = *'ISO 2022 escape sequences'*

comments character sets = *'ISO 2022 escape sequences'*

If the attribute **profile character sets** is missing, the character set consists of the 73 graphic characters of the minimum subrepertoire of ISO 6937, Part 2, with the additional control functions SPACE, LINE FEED and CARRIAGE RETURN.

4.2.6   Relations to a Document Application Profile

If the document conforms to a particular document application profile –
this is indicated by the document profile attribute document application
profile – the document application profile may specify default values for
a set of attributes which are different from those defined in ISO 8613.

Consider the following example: The parameter nominal page size of
the attribute medium type is used to indicate the physical size of a sheet
of paper on which the pages of an ODA document shall be printed. If a
value for this parameter is not specified explicitly in the document, the
default value defined in the Standard is the size of an ISO A4 page (see
p. 135). If, however, the document conforms to a document application
profile, a different default value for the page size may be defined by the
document application profile, for instance, the North American Letter
size.

To ensure that those specifications in a document application profile
which are required for the processing of the document can be taken into
account, these specifications are included in the document profile. This
is done by means of the attributes described in this section.

The value of the attribute document application profile defaults is a list
of the default values which are defined in the document application profile
to which the document conforms:

```
document application profile defaults =
   {[default value for block alignment]
    [default value for border]
    [default value for colour]
    [default value for content architecture class]
    [default value for dimensions]
    [default value for layout path]
    [default value for medium type]
    [default value for page position]
    [default value for transparency]
    [default value for type of coding]
    [default values for further presentation attributes and coding
     attributes of particular content architectures]}
```

As can be seen, the specification of default values in document application profile is not permitted for all attributes but only for block alignment,
border, colour, content architecture class, dimensions, layout path, medium
type, page position, transparency, type of coding and for several presen-

tation attributes and coding attributes which are defined in Parts 6, 7 and 8 of the Standard.

The detailed format of how the default values have to be specified shall not be given here. It depends on the kind of value permitted for an attribute and can be derived from the descriptions of the respective attributes. For instance, the value of the attribute **page position** is a sequence of non-negative integers (see p. 134) and a specification of a default value in the document profile would therefore be given in the form:

document application profile defaults = {**page position** = ⟦100 120⟧}

Of course, a specification of default values by the attribute **document application profile defaults** is only necessary, if the default values in the document application profile are different from those defined in ISO 8613.

Furthermore, certain attribute values can be classified as *basic* in a document application profile (see p. 25). If in a document conforming to a document application profile *non-basic* attribute values are used, this has to be explicitly announced in the document profile. Non-basic values are only possible for the attributes **block alignment, border, coding attributes, colour, dimensions, fill order, layout path, medium type, page position, presentation attributes, protection, transparency** and **type of coding**; the values of the other attributes are always basic.

The announcement of the usage of non-basic values in the document is made by means of several attributes in the document profile whose value is a list of the non-basic values of the respective attributes. The attributes of the document profile and their values are the following:

block alignments = $\{['$non-basic value for **block alignment**$']^+\}$

borders = $\{['$non-basic value for **border**$']^+\}$

coding attributes = $\{['$non-basic value for **coding attributes**$']^+\}$

colours = $\{['$non-basic value for **colour**$']^+\}$

fill orders = $\{['$non-basic value for **fill order**$']^+\}$

layout paths = $\{['$non-basic value for **layout path**$']^+\}$

medium types = $\{['$non-basic value for **medium type**$']^+\}$

page dimensions = $\{['$non-basic value for **dimensions**$']^+\}$

page positions = $\{['$non-basic value for **page position**$']^+\}$

presentation features =
  $\{[\text{'non-basic value for presentation attributes'}]^+\}$

protections = $\{[\text{'non-basic value for protection'}]^+\}$

transparencies = $\{[\text{'non-basic value for transparency'}]^+\}$

types of coding = $\{[\text{'non-basic value for type of coding'}]^+\}$

As can be seen, most of the names of document profile attributes are formed by adding an "s" to the name of the respective document body attribute. An exception are the attributes presentation features and page dimensions, where the document body attribute already had a plural form. A non-basic value for the attribute dimensions is only possible if the attributes is specified for a page which is explicitly expressed in the name of the document profile attribute.

The detailed format of how the non-basic values have to be specified shall not be given here. It depends on the kind of value permitted for an attribute and can be derived from the descriptions of the respective attributes. For example, a value of right-hand aligned may have been declared as the basic value for the attribute block alignment in a document application profile. If in a document conforming to this document application profile the values left-hand aligned and centred are also used, this has to be announced in the document profile by the specification

block alignments={'left-hand aligned' 'centred'}

The purpose of the announcement of non-basic values in the document profile should be obvious: a receiving system can decide by an analysis of the document profile whether it is able to process the document, in particular, whether it supports the features that were classified as non-basic in the document application profile but appear in a particular document.

4.2.7   Content Related Specifications

Several attributes of the document profile relate to the content of the document. These are the attributes title, subject, abstract, document type, keywords, languages, user-specific codes, document size and number of pages.

The attribute title gives the name of the document, the attribute subject specifies its subject, the attribute abstract contains a summary of the document and the attribute document type indicates its type such as "letter" or "internal report". The values of these attributes are *profile character set strings*:

title = '*profile character set string*'

subject = '*profile character set string*'

abstract = '*profile character set string*'

document type = '*profile character set string*'

The values of the attributes keywords, languages and user-specific codes consist of one or more entries where each entry assigns a keyword to the document (keywords), lists one of the language used (languages) or indicates additional information such as a contract number or budget code (user-specific codes). Each entry is a *profile character set string*:

keywords = $\{[\,'\textit{profile character set string}\,']^{+}\}$

languages = $\{[\,'\textit{profile character set string}\,']^{+}\}$

user-specific codes = $\{[\,'\textit{profile character set string}\,']^{+}\}$

The attribute document size represents the estimated size of the document including the document profile, expressed as the number of 8-bit bytes, where the estimate must not be less than the actual size. The attribute number of pages indicates the number of pages in the specific layout structure, if present. The values of these two attributes are positive or non-negative numbers:

document size = '*positive integer*'

number of pages = '*non-negative integer*'

All specifications made by these attributes are not needed for the editing, layout or imaging process and are primarily intended for evaluation by human readers. An automatic processing of some of these attributes is conceivable, however. For instance, an attribute such as keywords may be used for the retrieval of ODA documents in data bases.

4.2.8   Time Related Specifications

Several attributes of the document specify dates and times concerning
the history or future of the document. The values of these attributes are
always one or more calendar dates or times of day in a format defined
by ISO 8601 (see p. 183). The specification of the time is optional; the
calendar date is sufficient.   These document profile attributes are the
following:

creation date and time = '*date and time*'

document date and time = '*date and time*'

expiry date and time = '*date and time*'

local filing date and time = $\left[\left[\text{'}\textit{date and time}\text{'} \right]^+ \right]$

purge date and time = '*date and time*'

release date and time = '*date and time*'

start date and time = '*date and time*'

The attributes provide the following information:

The attribute creation date and time indicates when the document was
initially created.

The attribute document date and time gives a date which was assigned
by the originator of the document. (More details on the semantics of the
attribute are not given in the Standard.)

The attribute expiry date and time specifies when the document is
considered to be invalid.

The attribute local filing date and time indicates when the document
was filed on a storage medium. More than one entry may be given for
this attribute where the last one always indicates the most recent date
and, optionally, time of day.

The attribute purge date and time specifies when the document can be
purged from wherever it is stored.

The attribute release date and time gives a date and, optionally, time
of day after which the document can be released from any restrictions
defined by the attribute security classification (see p. 199).

The attribute start date and time specifies when the document is considered to be valid.

4.2.9   Specifications of the Development of the Document

In addition to the date specifications of the preceding section, the attributes organizations, preparers, owners, authors, status, copyright, authorization and revision history provide further information related to the development of the document.

The value of the attribute organizations is a set of one or more *profile character set strings* (see p. 182); each character string indicates an organization involved in the development of the document:

organizations $= \{['$ *profile character set string* $']^+\}$

The value of the attribute preparers is a set of one or more entries where each entry consists of the two parameters personal name of preparer and preparer's organization. The value of the parameter personal name of preparer is a *personal name* (see p. 182), the value of the parameter preparer's organization a *profile character set string*:

preparers $=$
$\quad \{[[$personal name of preparer $= '$ *personal name* $']$
$\quad [$preparer's organization $= '$ *profile character set string* $']]^+\}$

Each entry identifies one of the persons responsible for the physical preparation of the document. The parameters are both optional, but at least one parameter should be specified.

The value of the attribute owners is a set of one or more entries where each entry consists of the two parameters personal name of owner and owner's organization. The value of the parameter personal name of owner is a *personal name*, the value of the parameter owner's organization a *profile character set string*:

owners $=$
$\quad \{[[$personal name of owner $= '$ *personal name* $']$
$\quad [$owner's organization $= '$ *profile character set string* $']]^+\}$

Each entry identifies one of the persons responsible for the content of the document. The parameters are both optional, but at least one parameter should be specified.

The value of the attribute **authors** is a set of one or more entries where each entry consists of the two parameters **personal name of author** and **author's organization**. The value of the parameter **personal name of author** is a *personal name*, the value of the parameter **author's organization** a *profile character set string*:

> **authors** =
>   { [[**personal name of author** = '*personal name*']
>   [**author's organization** = '*profile character set string*']]$^+$}

Each entry identifies one of the persons responsible for the intellectual content of the document. The parameters are both optional, but at least one parameter should be specified.

The value of the attribute **status** is a *profile character set string* indicating the status of the document, e.g., "working paper" or "final version":

> **status** = '*profile character set string*'

The value of the attribute **copyright** is a set of one or more entries where each entry consists of the two parameters **copyright information** and **copyright dates**. The value of the parameter **copyright information** is a set of one or more *profile character set strings* identifying the names of the legal parties in whom the copyright of the document is vested. The value of the parameter **copyright dates** is a set of one or more calendar dates according to ISO 8601 which are associated with the copyright of the holders:

> **copyright** =
>   { [[**copyright information** = {['*profile character set string*']$^+$}]
>   [**copyright dates** = {['*date and time*']$^+$}]]$^+$}

The parameters of each entry are both optional, but at least one parameter should be specified.

The attribute **authorization** denotes the person or organization which approved or authorized the document. Its value is either a *personal name* or a *profile character set string*:

> **authorization** = ('*personal name*' | '*profile character set string*')

The attribute revision history describes the history of the document, indicating when, where and by whom the document was created or revised. The value of the attribute is a sequence of entries, where each entry consists of one or more of the parameters revision date and time, version number, reviser(s), version reference and user comments.

The parameter revision date and time indicates when a revision occurred; its value is a *date and time* according to ISO 8601. The value of the parameter version number is a *profile character set string* which assigns a version number to the revision. The value of the parameter reviser(s) is a set of one or more entries where each entry consists of the optional sub-parameters name(s), position and organization whose values are *personal names* or *profile character set strings* denoting the name, position and organization of a person who carried out the revision.

The value of the parameter version reference is either a *ASN.1 object identifier* or a *profile character set string*, indicating a source document which was the basis for the revised document. The value of the parameter is the same as the value of the attribute document reference in the profile of the source document (see p. 198). The value of the parameter user comments is a *profile character set string* which describes the revisions made. This can be summarized as follows:

revision history $=$
$\big[$ [[revision date and time $=$ '*date and time*']
   [version number $=$ '*profile character set string*']
   [reviser(s) $=$ {[[name(s) $=$ {['*personal name*']$^+$}]
                     [position $=$ '*profile character set string*']
                     [organization $=$ '*profile character set string*']]$^+$}
   [version reference $=$
        ('*ASN.1 object identifier*' | '*profile character set string*')]
   [user comments $=$ '*profile character set string*']]]$^+$ $\big]$

4.2.10   Relations to Other Documents

The attributes document reference, superseded documents and references to other documents can be used to establish relations between different ODA documents.

The value of these attributes consists of one entry (for document reference) or a set of entries (for the other two attributes) where each entry is either an *ASN.1 object identifier* (see p. 209) or a *profile character set string*:

document reference =
    ('*ASN.1 object identifier*' | '*profile character set string*')

superseded documents =
    $\{ [('ASN.1\ object\ identifier' | 'profile\ character\ set\ string')]^{+} \}$

references to other documents =
    $\{ [('ASN.1\ object\ identifier' | 'profile\ character\ set\ string')]^{+} \}$

The attribute document reference can be regarded as the "name" of the document; other documents can refer to it by means of this name, for instance, when the document is referenced with the attributes resource document, external document class or superseded documents.

The entries for the attribute superseded documents identify those documents which are superseded by the document for which the attribute is specified. Similarly, the entries of the attribute references to other documents list those documents which have some relation to the document. All these entries are the values of the attributes document reference of the referenced documents.

4.2.11   Access to the Document

The attributes distribution list, local file references, access rights and security classification contain information concerning the access of the document.

The value of the attribute local file references consists of one or more entries which indicate where a copy of the document can be found on a storage medium. Each entry consists of up to three parameters with the names file name, location of the document and user comments whose values are *profile character set strings*:

local file references =
    $\{ [[$file name = '*profile character set string*']
    [location of the document = '*profile character set string*']
    [user comments = '*profile character set string*'] $]^{+} \}$

The parameter file name identifies the document uniquely in a filing system. The parameter location of the document may provide further information where the document can be found in the filing system, for instance, the name of a directory, and additional comments on its location may be given with the parameter user comments.

The value of the attribute distribution list consists of one or more entries which indicate the intended recipients of the document. Each entry consists of the two parameters personal name of recipient and recipient(s) organization; the value of the parameter personal name of recipient is a *personal name* and the value of the parameter recipient(s) organization a *profile character set string*:

distribution list =
   $\big\{$ [[personal name of recipient = '*personal name*']
     [recipient(s) organization = '*profile character set string*'] $]^+\big\}$

Both parameters are optional but at least one parameter should be specified for each entry.

The value of the attribute access rights consists of one or more entries. Each entry is a *profile character set string* which specifies the access rights to the document as defined by the current owner of the document:

access rights = $\big\{$ ['*profile character set string*'$]^+\big\}$

The value of the attribute security classification is a *profile character set string* specifying the security classification assigned by the owner relating to such aspects as visibility, reproduction, storage, audit and destruction requirements:

security classification = '*profile character set string*'

4.2.12   Security Attributes

The seven attributes ODA security label, sealed document profiles, pre-sealed document body parts, post-sealed document body parts, enciphered document profiles, pre-enciphered document body parts and post-enciphered document body parts were introduced by the Addendum on Security, in particular, to deal with the protected document part specified by this Addendum (see also Sect. 3.5).

The concept behind these attributes is as follows. For each constituent in the protected part of an ODA document, i.e., for a sealed document profile, enciphered document profile, pre-enciphered document body part and post-enciphered document body part, a particular encryption method has been applied in order to prevent an unauthorized receiver of the document from getting semantic knowledge of the content of these constituents. A privileged receiver may use the document profile attributes enciphered document profiles, pre-enciphered document body parts and post-enciphered document body parts to decipher these constituents, however, additional information – only known to a privileged receiver – is required for this process.

Furthermore, seals may be associated with the document as a whole and constituents of the document by means of the attributes ODA security label, sealed document profiles, pre-sealed document body parts and post-sealed document body parts. Such seals may be used for additional security issues (see Sect. 3.5).

It is also possible to distinguish between several privileged receivers where each receiver may have different access rights to constituents of the protected part of a document. For instance, a constituent X may only be decipherable for receiver A and a constituent Y only for receiver B.

The attribute ODA security label has the two parameters ODA label and ODA label seal. The value of the parameter ODA label is a character string from the document profile character set. The value of the parameter ODA label seal is an octet string:

ODA security label =
    {ODA label = '*profile character set string*'
      ODA label seal = '*octet string*'}

This attribute defines a security label associated with the document. The octet string of the parameter ODA label seal represents the seal which is derived from the value of the parameter ODA label and the value of the attribute document reference.

The value of the attribute sealed document profiles is a sequence of entries where each entry consists of the parameters document profile seal, privileged recipients and sealed document profile identifier. The parameter document profile seal has the sub-parameters seal method, sealed information and seal.

The sub-parameter seal method has the sub-sub-parameters fingerprint information, fingerprint method, sealing information and sealing method.

The value of the sub-sub-parameters fingerprint information and sealing information is a *key information* (see p. 183). The value of the sub-sub-parameters fingerprint method and sealing method is sequence of an ASN.1 object identifier and a character string from the document profile character set.

The sub-parameter sealed information has the sub-sub-parameters fingerprint, location, seal originator and time. The value of the sub-sub-parameter fingerprint is an octet string, the value of the sub-sub-parameter location is sequence of an ASN.1 object identifier and a character string from the document profile character set, the value of the sub-sub-parameter seal originator is a personal name and the value of the sub-sub-parameter time is a date and time value according to the format defined in ISO 8601. The value of the sub-parameter seal is an octet string.

The value of the parameter privileged recipients is a set of *personal names*. The value of the parameter sealed document profile identifier is a *protected-part-id*. We therefore have:

sealed document profiles =
  { [document profile seal =
    {seal method =
      {fingerprint information = '*key information*'
      fingerprint method = ['*ASN.1 object identifier*'
                      '*profile character set string*']
      sealing information = '*key information*'
      sealing method = ['*ASN.1 object identifier*'
                 '*profile character set string*']}
    sealed information =
      {fingerprint = '*octet string*'
      location = ['*ASN.1 object identifier*'
              '*profile character set string*']
      seal originator = '*personal name*'
      time = '*date and time*'}
    seal = '*octet string*'}
  privileged recipients = {['*personal name*']$^+$}
  sealed document profile identifier = '*protected-part-id*']$^+$}

Each entry of the attribute value specifies information associated with each sealed document profile. The parameter privileged recipients lists the names of the authorized receivers of the document for that sealed document profile which is identified by the parameter sealed document profile identifier. The algorithm by which the seal was created is defined by the sub-sub-parameters of the sub-parameter seal method. The seal itself is given by the sub-parameter seal. The sub-sub-parameters of the sub-parameter sealed information describe where (location), when (time) and by whom (seal originator) the seal was created and the fingerprint (see p. 158) used to create the seal.

The value of the attribute pre-sealed document body parts is a sequence of entries where each entry consists of the parameters document body part seal, privileged recipients, sealed constituents and seal identifier. The parameter document profile seal has the same structure as the corresponding parameter of the attribute sealed document profiles.

The value of the parameter privileged recipients is a set of *personal names*. The value of the parameter sealed constituents is a a sequence of *constituent identifiers* (see p. 72) and the value of the parameter seal identifier is an integer:

$$
\begin{aligned}
&\text{pre-sealed document body parts} = \\
&\quad \{\,[\text{document body part seal} = \\
&\qquad \{\text{seal method} = \\
&\qquad\quad \{\text{fingerprint information} = \text{'}key\ information\text{'} \\
&\qquad\quad\ \ \text{fingerprint method} = [\![\ \text{'}ASN.1\ object\ identifier\text{'} \\
&\qquad\qquad\qquad\qquad\qquad\qquad\ \ \text{'}profile\ character\ set\ string\text{'}\ ]\!] \\
&\qquad\quad\ \ \text{sealing information} = \text{'}key\ information\text{'} \\
&\qquad\quad\ \ \text{sealing method} = [\![\ \text{'}ASN.1\ object\ identifier\text{'} \\
&\qquad\qquad\qquad\qquad\qquad\qquad\ \ \text{'}profile\ character\ set\ string\text{'}\ ]\!]\} \\
&\qquad \text{sealed information} = \\
&\qquad\quad \{\text{fingerprint} = \text{'}octet\ string\text{'} \\
&\qquad\quad\ \ \text{location} = [\![\ \text{'}ASN.1\ object\ identifier\text{'} \\
&\qquad\qquad\qquad\qquad\ \ \text{'}profile\ character\ set\ string\text{'}\ ]\!] \\
&\qquad\quad\ \ \text{seal originator} = \text{'}personal\ name\text{'} \\
&\qquad\quad\ \ \text{time} = \text{'}date\ and\ time\text{'}\} \\
&\qquad \text{seal} = \text{'}octet\ string\text{'}\} \\
&\quad \text{privileged recipients} = \{[\text{'}personal\ name\text{'}]^+\} \\
&\quad \text{sealed constituents} = [\![\ [\text{'}constituent\ identifier\text{'}]^+\ ]\!] \\
&\quad \text{seal identifier} = integer]^+\}
\end{aligned}
$$

Each entry of the attribute value specifies information associated with each pre-sealed document body part (see Sect. 3.5). The parameters document body part seal and privileged recipients have the same semantics as for the attribute sealed document profiles. The parameter sealed constituents lists the constituents of the document which were used to create the seal. The value of the parameter seal identifier is a "name" associated with a particular seal. This integer appears also in the value of the parameter seal identifiers of the attribute sealed (see p. 136).

The attribute post-sealed document body parts has the same structure as the attribute pre-sealed document body parts:

post-sealed document body parts =
  {[document body part seal =
    {seal method =
      {fingerprint information = '*key information*'
       fingerprint method = [['*ASN.1 object identifier*'
                           '*profile character set string*']]
       sealing information = '*key information*'
       sealing method = [['*ASN.1 object identifier*'
                    '*profile character set string*']]}
     sealed information =
      {fingerprint = '*octet string*'
      location = [['*ASN.1 object identifier*'
               '*profile character set string*']]
      seal originator = '*personal name*'
      time = '*date and time*'}
    seal = '*octet string*'}
  privileged recipients = {['*personal name*']$^+$}
  sealed constituents = [['*constituent identifier*']$^+$]]
  seal identifier = *integer*]$^+$}

Each entry of the attribute value specifies information associated with each post-sealed document body part (see Sect. 3.5). The meaning of the parameters and sub-parameters is the same as for the attribute pre-sealed document body parts.

The value of the attribute enciphered document profiles is a sequence of elements where each element consists of the parameters protected document part identifier and privileged recipient information. The value of the parameter protected document part identifier is a *protected-part-id*. The parameter privileged recipient information has the sub-parameters key information, method information and privileged recipients. The value of the sub-parameter privileged recipients is a set of *personal names* and the value of the sub-parameter method information is an ASN.1 object identifier followed by a character string from the document profile character set. The value of the sub-parameter key information is a *key information* (see p. 183):

```
enciphered document profiles =
  {[privileged recipient information =
     {key information = 'key information'
      method information = [['ASN.1 object identifier'
                              'profile character set string']]
      privileged recipients = {['personal name']⁺}}
   protected document part identifier = 'protected-part-id']⁺}
```

Each entry of the attribute value specifies information which is required to decipher the enciphered document profile identified by the parameter **protected document part identifier**. The sub-parameter **privileged recipients** lists the persons authorized to decipher this constituent. The sub-parameter **method information** specifies the method which was used to encipher the constituent. The sub-parameter **key information** provides information for a privileged recipient on how to deduce the key needed to decipher the enciphered document profile.

The attributes **pre-enciphered document body parts** and **post-enciphered document body parts** have the same structure as the attribute **enciphered document profiles**:

```
pre-enciphered document body parts =
  {[privileged recipient information =
     {key information = 'key information'
      method information = [['ASN.1 object identifier'
                              'profile character set string']]
      privileged recipients = {['personal name']⁺}}
   protected document part identifier = 'protected-part-id']⁺}

post-enciphered document body parts =
  {[privileged recipient information =
     {key information = 'key information'
      method information = [['ASN.1 object identifier'
                              'profile character set string']]
      privileged recipients = {['personal name']⁺}}
   protected document part identifier = 'protected-part-id']⁺}
```

Each entry of the attribute values specifies information which is required to decipher the pre-enciphered or post-enciphered document body part identified by the parameter **protected document part identifier**. The meaning of the parameters, sub-parameters and sub-sub-parameters is the same as for the attribute **enciphered document profiles**.

4.2.13   Additional Specifications

There are three attributes of the document profile left that were not described in the preceding sections: unit scaling, fonts list and additional information.

Many constituents in ODA documents may contain attributes which specify absolute or relative positions or dimensions measured in so-called *scaled measurement units*. These scaled measurement units are transformed into basic measurement units using a scaling factor (see Sect. 3.2.18). The scaling factor is defined by the document profile attribute unit scaling. Its value consists of two positive integers:

unit scaling $= [\![\,$ '*positive integer* ' '*positive integer* ' $]\!]$

If the first integer is called m and the second n, one scaled measurement unit is defined as m/n basic measurement unit. It should be noted that this attribute is the only document profile attribute for which a default value is specified in the Standard: if the attribute is missing, the values $m = n = 1$ are assumed, i.e., one scaled measurement unit is equal to one basic measurement unit.

Part 6 of the ODA Standard (Character Content Architectures) defines several attributes to select different fonts (graphic character sets) within a text. Within the document body, these fonts are identified by numbers; the assignment of an actual font to a particular number is performed by the document profile attribute fonts list. The value of the attribute consists of one ore more entries where each entry is composed of a number and a so-called *font reference* according to ISO 9541, Part 5 (see p. 183):

fonts list $= \big[\, \big[integer$ '*font reference* ' $\big]^{+} \,\big]$

Finally, there exists the document profile attribute additional information, which can have any value and whose semantics are not explicitly defined in the Standard.

5 Part 5: Office Document Interchange Format (ODIF)

The ODA Standard does not define a data format which has to be used by an ODA implementation for the internal storage of documents; such an internal format is considered implementation dependent.

When, however, an ODA document is to be electronically interchanged, a precisely specified coding has to be used to guarantee that a receiving system can decode the sequence of bits and transform it into its internal format. This coding is specified in Part 5 of ISO 8613, i.e., this Part defines the precise rules for the data stream of an interchanged document.

Part 5 specifies two different encodings. The one encoding – called ODIF – is based on the *Abstract Syntax Notation One* (ASN.1) as defined in the ISO Standards 8824 and 8825. ASN.1 encodings are also used in many other ISO Standards and CCITT Recommendations in the area of *Open Systems Interconnection* (OSI). It is a binary encoding, i.e., an ASN.1 data stream (and therefore also an ODIF data stream) is not intended to be read by humans but it should rather be decoded by software.

The second encoding – called *Office Document Language* (ODL) – is a *clear text encoding*, i.e., it can also be read by humans, though this would usually be an exception. The ODL encoding is based on the *Standard Generalized Markup Language* (SGML). SGML is also an ISO Standard with the Registration Number 8879. (More precisely, ODL is not directly used for the interchange of ODA documents but the encoding of ODL documents in the *SGML Document Interchange Format (SDIF)* as defined in ISO 9069 but this is only a minor detail which shall be ignored.)

The ODL encoding scheme was included in the ODA Standard to build a bridge between the "ODA world" and "SGML world". It provides a means of encoding an ODA document in such a way that it can be regarded as an SGML document. The opposite direction – transforming an SGML document in such a way that it can be regarded as an ODA document – is usually not possible; however, the problems involved shall not be discussed here.

The ODL encoding is the only essential difference between ISO 8613 and the CCITT T.410 series of Recommendations: ODL is a normative part of ISO 8613 but it is not included in the T.410 documents. One reason for this difference is the fact that SGML is not a CCITT Recommendation and is rather unlikely to become one because applications of the SGML Standard are slightly outside the scope of CCITT.

Since ODL is located at the edge of the ODA Standard and since major applications of ODL are currently not known it shall not be discussed in detail in this book.

Furthermore, ODIF itself will also not be described in greater depth. The technical specifications of Part 5 consist essentially of detailed encoding rules in a formal notation which are of rather limited importance for the understanding of the general principles of the Standard. Only when an ODA system is implemented these encoding rules have to be studied, then, however, very thoroughly.

5.1   The Abstract Syntax Notation One (ASN.1)

As mentioned above, the ODIF encoding of ODA documents is based on the ISO Standards 8824 (*Open Systems Interconnection – Specification of Abstract Syntax Notation One*) and 8825 (*Open Systems Interconnection – Specification of basic encoding rules for Abstract Syntax Notation One*).

ASN.1 is a notation which provides for the application-dependent definition of data types, starting from a set of elementary data types such as integers or octet strings. The methodology permits the creation of very complex data types up to the complete specification of the syntax of a binary data stream which is to be exchanged between two systems. ASN.1 can both be used for

- the definition of the syntax, i.e., of the rules according to which a data stream is built, and
- the data stream itself.

In the area of *Open Systems Interconnection* – and also in the ODA Standard – these two features of ASN.1 are usually applied in such a way that a Standard defines abstract syntax rules for the data streams to be interchanged, and a particular implementation of the Standard can create and receive data streams conforming to these syntax rules.

The ASN.1 Standards (ISO 8824 and 8825) already contain the definition and encoding rules for a set of elementary data types such as *boolean*, *integer*, *real*, *bit string*, *octet string*, *set*, *sequence* or *object identifier*.

An *ASN.1 object identifier* is a sequence of non-negative integers. The main purpose of these identifiers is to provide a unique naming scheme for "objects" in the CCITT and ISO world. For instance, the attribute content architecture class may have the ASN.1 object identifier '2 8 2 8 0' as its value (see p. 305) to identify the content type of a content portion as geometric graphics content. Any CCITT Recommendation or ISO Standard may use this object identifier only with this meaning, i.e., the semantics of this identifier are defined by the ODA Standard for the whole CCITT and ISO world. A joint registration authority of CCITT and ISO takes care of the uniqueness of the identifiers.

ASN.1 object identifiers are used frequently as data types for values of attributes or parameters in the ODA Standard. Sometimes the identifiers themselves are already specified in the Standard, for example, for the attributes content architecture class or type of coding, but for a number of other attributes the actual identifiers are not defined in the Standard.

For instance, the document profile attribute document reference may have an ASN.1 object identifier as its value (see p. 198) which is the "name" of a particular ODA document by which it may be referenced from other ODA documents. Obviously, the actual value of this attribute cannot be specified by the Standard. If an author wants to assign an object identifier to a document, this identifier must be approved by the registration authority. (Of course, this would not be very sensible for documents such as business letters with a rather limited distribution, but it may be useful, for example, for documents such as manuals of a computer system which are used by a large community to identify a particular version of a document uniquely.)

The specification of additional application-dependent data types is performed by so-called ASN.1 *definitions* whose syntax is based on elementary or other defined composite data types.

For a better understanding of these principles consider the following two examples. In Part 2 of the ODA Standard it is specified that the value of the attribute colour is either colourless or white. Concerning the ODIF encoding of this attribute value, Part 5 of the Standard contains the following ASN.1 definition:

Colour ::= INTEGER {colourless(0), white(1)}

This definition introduces a new data type with the name "Colour". This data type is represented as an integer – an elementary ASN.1 data type – whose value is either 0 or 1. The value 0 represents the attribute value colourless; the value 1 represents the attribute value white.

The attribute colour may be specified for presentation styles. Concerning the ODIF encoding of presentation styles, Part 5 contains the following (slightly simplified) ASN.1 definition:

Presentation-Style-Descriptor ::= SET{
    style-identifier   Style-Identifier,
    user-readable-comments   [0] Comment-String OPTIONAL,
    user-visible-name   [1] Comment-String OPTIONAL,
    transparency   [2] Transparency OPTIONAL,
    presentation-attributes   [3] Presentation-Attributes OPTIONAL,
    colour   [4] Colour OPTIONAL,
    border   [5] Border OPTIONAL,
    sealed   [6] Sealed OPTIONAL,
    derived-from   [7] Style-Identifier OPTIONAL}

This definition specifies that the ODIF encoding of a presentation style is based on the predefined ASN.1 data type *set*. This set may contain nine elements according to the nine attributes which may be specified for a presentation style (see p. 65). One of these elements is a data element of type "Colour". The element may be missing which is expressed by the ASN.1 keyword "OPTIONAL".

Within the set the data element of type "Colour" is identified by a so-called *tag* with the value 4. Of course, the other data types appearing in this definition – Style-Identifier, Comment-String, Transparency, Presentation-Attributes, Border and Sealed – are also defined in Part 5 of the Standard.

The major part of the rules for the ODIF encoding of a document consists of such ASN.1 definitions. Part 5 of the Standard contains the encoding rules for the constituents described in Part 2 and 4, i.e., for the objects, object classes, styles, content portions and the document profile; Parts 6, 7 and 8 contain the encoding rules for the different content architectures.

In the rest of this section the ODIF encoding of a concrete document according to the ASN.1 Standards and the additional specifications in the ODA Standard shall be outlined briefly.

Each ASN.1 data element, for instance, an integer or a composite data element, has the following structure:

identifier	length	contents	end of contents
octets	octets	octets	octets

An octet consists of eight bits. The identifier octets indicate the type of the data element, the length octets its length and the contents octets contain its actual value. The end of contents octets are only present, if the length octets specify an "indefinite" length.

In many cases one identifier octet is sufficient and, therefore, the case of more than one identifier octet shall not be considered in the following. The eight bits of an identifier octet are structured as follows:

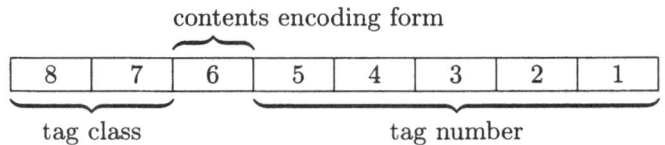

Bits 8 and 7 specifies the class of the data element. The ASN.1 Standard defines the following four classes:

00: *universal*
01: *application*
10: *context-specific*
11: *private*

If bit 6 has the value 0, the data element is *primitive*, for instance, an integer; if it has the value 1, it is a *constructed* data element, for example, a *set*.

The remaining five bits may have any bit combination except 11111; this bit combination indicates that more than one identifier octet is used for the specification of the data type. In case of a single identifier octet, the tag number can therefore be any number between 0 (binary: 00000) and 30 (binary: 11110).

For clarification consider the following examples. According to the ASN.1 Standard, an integer is a primitive data type which has the universal tag number 2. An integer has therefore the following identifier octet:

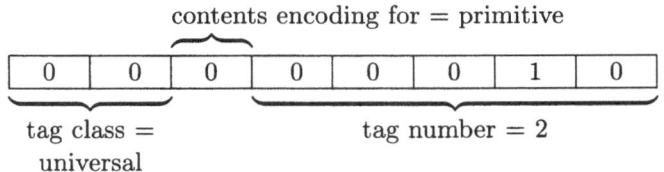

The ASN.1 encoding of the decimal number 1000 is therefore the following bit sequence:

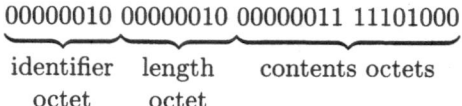

The identifier octet is the one defined in the ASN.1 Standard, the length octet has the value two, since two octets are needed for a binary encoding of the decimal number 1000, and the contents octets are as shown which is the usual binary representation of the number 1000.

The ASN.1 encoding of the attribute colour with the value **white** for a presentation style is – according to the ASN.1 definitions of "Presentation-Style-Descriptor" and "Colour" shown above – as follows:

$$10000100\ 00000001\ 00000001$$

identifier   length   contents
octet   octet   octet

The contents octet is the integer 1 (00000001), since the value **colour** is represented as the integer 1 according to the ASN.1 definition of "Colour" (see p. 209). The length octet has also the value 1 because only one content octet is needed. The bits 5 to 1 of the identifier octet are the number 4 (00100), since a data element of type "Colour" is identified by a tag number of 4 according to the ASN.1 definition of "Presentation-Style-Descriptor". Bit 6 has the value 0 because the value of the data element is an integer, i.e., primitive. Bits 8 and 7 are binary 10, since the tag number 4 is context specific: other data types may use this tag number for the identification of other data elements than "Colour" and, vice versa, a "Colour" data element may be identified by a tag number different from 4 in another context.

5.2   Additional Encoding Rules for ODA Documents

As explained in the preceding section, the ODIF encoding of an ODA document is a binary data stream according to a precisely defined ASN.1 syntax.

The ODA Standard distinguishes between the so-called *interchange format class A* and *interchange format class B*. The interchange format class A is permitted for all ODA documents whereas the interchange format class B can only be used for documents which contain only a generic or specific layout structure, presentation styles, content portions and a document profile.

For the interchange format class A, the different types of constituents have to appear in the following order in the ODIF data stream:

a)   document profile,
b)   layout object classes,
c)   logical object classes,
d)   content portions associated with object classes,
e)   presentation styles,
f)   layout styles,
g)   layout objects,
h)   logical objects,
i)   content portions associated with objects,
j)   sealed document profile descriptions,
k)   enciphered document profile descriptions,
l)   pre-enciphered document body parts and
m)   post-enciphered document body parts.

Of course, several of these groups of constituents may be missing in a particular data stream, for instance, layout objects in a document of the processable document architecture class.

The layout objects and the logical objects have to appear in their sequential order (see p. 91) in the data stream. If the document contains a specific layout structure, the content portions appear in their sequential layout order, otherwise, if only a specific logical structure is present, in their sequential logical order. Within the other groups, the order of the constituents in the data stream is arbitrary.

If alternative descriptions exist in the document (see Sect. 3.2.14), the basic objects representing such alternative descriptions follow in the data stream immediately after the basic objects representing the associated primary descriptions. If several alternative descriptions exist for a particular primary description, the respective basic objects follow in the order of decreasing preference. Likewise, the content portions for alternative descriptions follow in the data stream immediately after the content portions for the primary descriptions, also in decreasing preference order if several alternative descriptions exist.

The interchange format class B can be used for documents without a specific logical structure, generic logical structure or layout styles. The order in which the constituents have to appear in the data stream is as follows:

a) document profile,
b) layout object classes and associated content portions,
c) layout styles,
d) layout objects and associated content portions,
e) sealed document profile descriptions,
f) enciphered document profile descriptions and
g) post-enciphered document body parts.

Again, some of these groups may be missing in the data stream, for example, layout object classes and associated content portions.

The content portions associated with basic page classes or block classes – other layout object classes can have no associated content portions – follow in the data stream immediately after the object classes to which they belong. The layout objects appear in their sequential order in the data stream; content portions follow also immediately after the layout objects to which they belong.

It should be noted that the hierarchical tree structure of the specific layout objects and their associated content portions can be immediately derived from the order in which they appear in an ODIF data stream. Therefore, when using the interchange format class B, the attributes object identifier, content identifier layout, subordinates and content portions which usually are needed for defining this structure are not required (see Sect. 3.3.3).

6 Part 6: Character Content Architectures

This Part of the Standard defines the encoding of character content in ODA documents, specifies the attributes related to character content and describes how the layout process and imaging process for character content has to be carried out. For simplicity, the term "text" shall also be used in the following to denote character content.

6.1 The ODA Model for Character Content

The ODA model for text comprises not only Latin scripts, i.e., scripts with a writing direction from left to right (and top to bottom), but it is also able to handle other kinds of scripts such as Arabic or Kanji where the writing direction is from right to left or top to bottom. This explains why several of the features below which are of no importance for Latin scripts have been included in the Standard.

On the other hand, the ODA model for character content does not provide for the processing of all kinds of text. For instance, more complex mathematical formulae or highly structured tables can hardly be handled by this model. However, extensions of the Standard into such areas may be expected in the future.

In the following, all features of the ODA text model will be explained, but the emphasis will be on Latin scripts.

6.1.1 Characters

The basic elements in the ODA model for text are *graphic characters* which are members of a set of graphic symbols used for the representation of information. The rendition of a graphic character on a presentation medium is called a *character image*. Each character image has an associ-

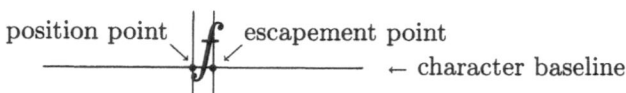

Fig. 44: Character baseline, position point and escapement point of a character image

ated *character baseline*, a *position point* and *escapement point* as shown in Fig. 44.

The character baseline is an imaginary line through the character image to define the orientation of the character: For a horizontal baseline and a *character orientation* (see below) of 0° the character image is in its intended viewing orientation.

The *position point* is a reference point associated with the character image and is used for positioning the character image within a *line box* (see below). The *escapement point* is another reference point of the character image: the distance between position point and escapement point defines the width of the character. For two consecutive characters, the position point of the second character is placed on the escapement point of the first character unless an explicit character distance is specified (see p. 219) or *kerning* takes place (see p. 222).

It should be noted that a character image may extend beyond its position or escapement point. Such an extension is called a *kern*.

Consecutive characters are assembled within a line box in the direction of the *character path*. The character images have a *character orientation* which is defined by the counterclockwise angle between character path and character baseline as shown in Fig. 45.

The position point and escapement point are labeled with "pp" and "ep" in the figure. It should be noted that character path and character baseline need not necessarily be parallel to each other.

Each character belongs to a particular *font*. A font is a set of character images which share characteristics such as a common design and size. Each font has a set of properties which are defined by attributes. Also the properties of a particular character of a font are usually defined by attributes. These attributes, however, are not defined in the ODA Standard but they are taken from the ISO Standard 9541, Part 1, (*Font and character information interchange – Part 1: Architecture*).

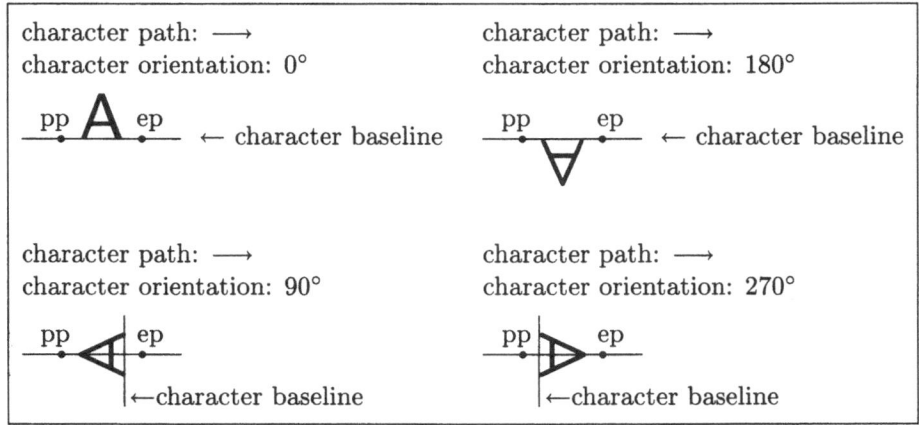

character path: ⟶
character orientation: 0°

pp A ep ← character baseline

character path: ⟶
character orientation: 180°

pp   ep ← character baseline

character path: ⟶
character orientation: 90°

pp   ep
      ←character baseline

character path: ⟶
character orientation: 270°

pp   ep
      ←character baseline

Fig. 45: Character orientation, character path and character baseline

6.1.2 Positioning of Characters

The individual characters of a text are displayed within the *positioning area*. This a rectangular area completely contained within the area of the basic layout object to which the text belongs. The four edges of the positioning area are called *start edge, end edge, top edge* and *bottom edge*, depending on the direction of the character path and line progression as shown in Fig. 46.

The figure shows that the start edge and end edge of the positioning area need not necessarily be coincident with the respective edges of the basic layout object but they may be a certain distance apart. Such a distance is called a *kerning offset*. Parts of the character images may extend into the area of the kerning offset (see Fig. 44); however, no parts of the character images may extend beyond the boundary of the basic layout object.

Within the positioning area, a sequence of character images is positioned within a rectangular area called a *line box* which extends from the start edge to the end edge of the positioning area as shown in Fig. 47.

As can also be seen in this figure, each line box contains an imaginary line called the *reference line* which is used for the alignment of the character images. The *line home position* indicates that point on the line home position where usually the position point of the first character image in the line is aligned. The distance between the start edge of the positioning area and the line home position is called the *indentation.*

The length of a line box equals the distance between end edge and start edge of the positioning area; its height is the sum of the *backward*

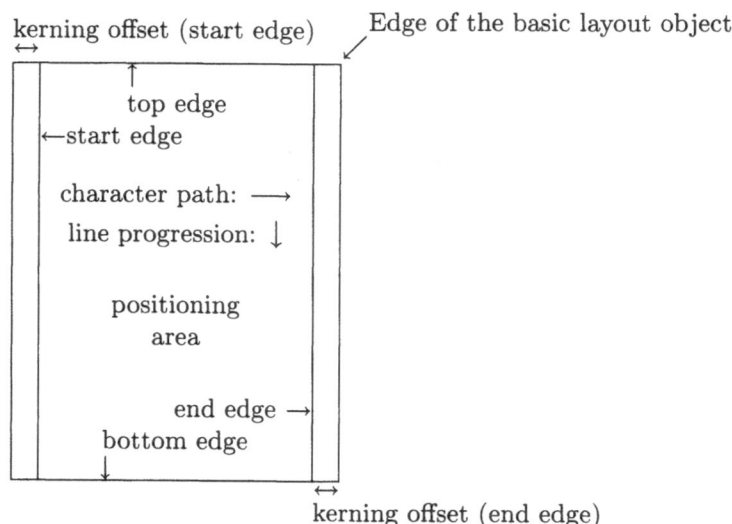

Fig. 46: Edges of the positioning area in relation to the character path and line progression

extent and *forward extent*. The sizes of the backward extent and forward extent are defined by the fonts from which the characters in the line were chosen. In particular, when chosing fonts conforming to ISO 9541, these sizes can be derived from certain attributes of the font. Of course, if some character images in the line are not aligned at the reference line, for instance, subscripts or superscripts, this has to be taken into account for the determination of the backward and forward extent.

Consecutive character images are aligned along the reference line of a line box. To describe some of the properties of this alignment process, the ODA Standard introduces the concept of the *active position*. The active position indicates an abstract point in the positioning area where the next "action" is going to take place during the formatting of a text, i.e., the active position is conceptually similar to a "cursor" which is usually visible during the editing of a text on a computer screen.

The next action during formatting is defined by the next element in the data stream. If the next element is a graphic character, its character image is positioned with the position point at the active position and the active position is advanced into the direction of the character path according to the spacing between the characters. If the next character is a control function (see Sect. 6.3), this may cause the active position to move to another point within the positioning area.

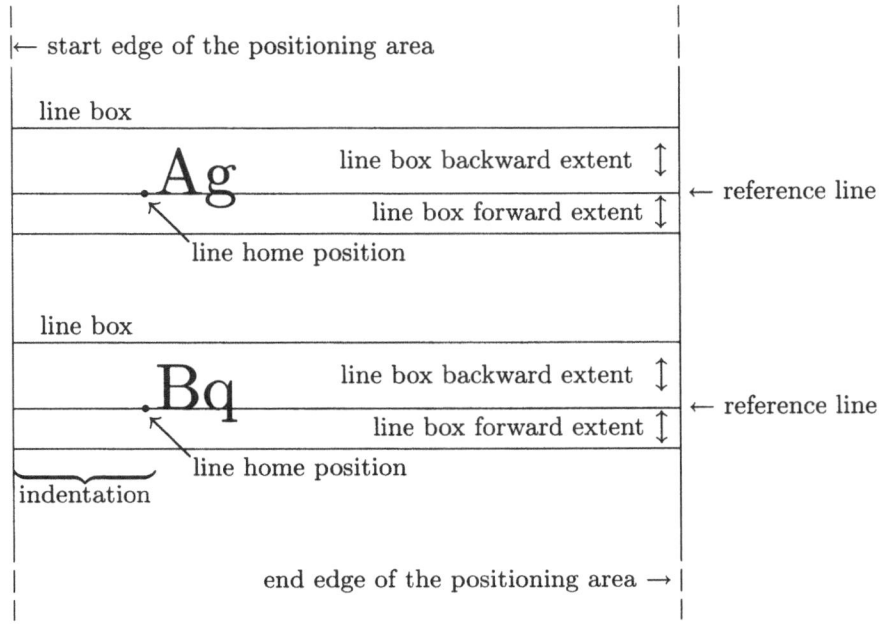

Fig. 47: Positioning of characters in line boxes

Spacing between characters

Concerning the distance between two consecutive characters, two different methods have to be distinguished: *constant character spacing* and *variable character spacing.*

For fonts with *constant spacing* the distance between the position points of consecutive character images is always the same; the distance can be considered a property of the font.

For fonts with *variable spacing* the distance between the position points of consecutive character images depends of the widths of the individual characters, i.e., the position point of a succeeding character is usually placed on the escapement point of the preceding character.

For both methods, the distance between consecutive characters may be increased by a *inter-character space* as shown in Fig. 48.

The left side of the figure shows a font with constant spacing and the right side with variable spacing. A positive inter-character space was used in both cases, i.e., the text runs wider than usually. The inter-character space may also be negative, i.e., a text would then run narrower than usually.

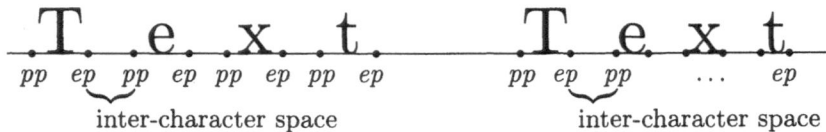

Fig. 48: Character spacing with inter-character space

A special role is played by the space character. For fonts with constant spacing its width equals the width of any other character in the font. For fonts with variable spacing its width is usually a property of a font which specifies a "nominal space width". This nominal width may be shrunk or stretched within a certain range to achieve justification during formatting (see next section).

In addition to the possibilities described above, the ODA Standard provides also several control functions to specify the character spacing (see p. 244).

Alignment within a line box

Concerning the alignment of character images within a line box, the ODA Standard defines four different methods:

- *start-aligned*: the position point of the first character image in each line box is placed at the line home position;
- *end-aligned*: the escapement point of the last character image in each line box is placed at the end edge of the positioning area;
- *centred*: the distance from the line home position to the position point of the first character image is equal to the distance of the escapement point of the last character image to the end edge of the positioning area;
- *justified*: the position point of the first character image is placed at the line home position and the escapement point of the last character image is placed at the end edge of the positioning area by chosing an appropriate space width and, if present, intercharacter space.

Besides these four justification methods when formatting a piece of text, additional features may influence the alignment of character images in a line box, for instance, the usage of tabulation stops.

Tabulation

It is possible to specify one or more *tabulation stops* within a line box. Each tabulation stop defines a point on the reference line of the line box relative to the start edge of the positioning area. A string of characters may be placed at a tabulation stop by means of a control function. One of the following four kinds of alignment at the tabulation stop is possible:

- *start-aligned*: the position point of the first character image of the character string is placed at the tabulation stop;
- *end-aligned*: the escapement point of the last character image of the character string is placed at the tabulation stop;
- *centred*: the character string is placed in such a way that the position point of the first character image has the same distance from the tabulation stop as the escapement point of the last character image;
- *aligned around*: an additional substring is specified and the position point of the first character image of this substring is positioned at the tabulation stop, if it appears within the character string. If the substring does not appear within the character string, the end-aligned tabulation method is applied.

These four methods of alignment are shown in Fig. 49.

This	This	This	126,48
is	is	is	54,50
start-	end-	centred	30183
aligned	aligned		21,985

Fig. 49: Alignment at tabulation stops

The vertical lines in the figure denote the position of the tabulation stops. The alignment at the first tabulation stop is start-aligned, at the second end-aligned and at the third centred. The forth tabulation stop shows the effect of aligned around where the substring consists only of one character, a comma. In fact, the alignment of decimal numbers is an important application of the aligned around method.

Subscripts and superscripts

The active position is usually on the reference line and therefore the position points of the character images are usually placed on that line. This can be changed, however, by means of several control functions (see Sect. 6.3) which move the active position in the direction or opposite to the direction of the line progression. This is needed, for instance, for the rendition of subscrips or superscripts.

Such a displacement of the active position from the reference line must always be cancelled before the end of a line box is reached, i.e., before a *hard line terminator* or *soft line terminator* occurs in the text (see p. 250). A detailed description of subscript and superscript rendition is given with the control functions PLD, PLU, VPB and VPR (see p. 243ff.).

Pairwise kerning

For certain combinations of characters, the general rule that the position point of the second character is positioned at the escapement point of the first character is not appropriate when a good typographical quality is required. Figure 50 shows the combination of the two characters "W" and "A" – on the left hand side when the position point of the "A" is placed at the escapement point of the "W", and on the right hand side when a so-called *pairwise kerning* is applied.

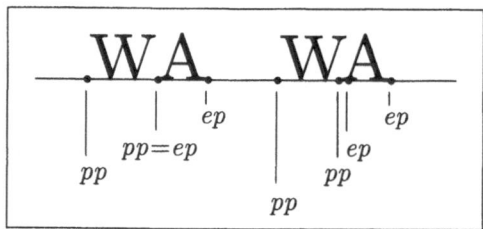

Fig. 50: Pairwise kerning for combinations of characters

The typographical result is obviously better if the position point of the "A" is placed slightly left of the escapement point of the "W" as shown in the right hand side of the figure; otherwise, the distance between the two characters appears to wide. Such an adjustment of the position point is called *pairwise kerning*. The displacement of the position point need not necessarily be opposite to the direction of the character path; for certain character combinations a shift into that direction is appropriate.

The specifications describing the pairwise kerning for character combinations – in particular, the necessary amount of shift – are a property of the font used and are defined by corresponding attributes of the font according to ISO 9541.

It should be noted that pairwise kerning can only take place if variable character spacing applies.

First line offset and itemization

Besides the indentation which effects a displacement of the line home position from the start edge of the positioning area as shown in Fig. 47, the ODA Standard provides two methods for a special treatment of the first line of a piece of text. The first method is the specification of a so-called *first line offset* which may be positive or negative as shown in Fig. 51.

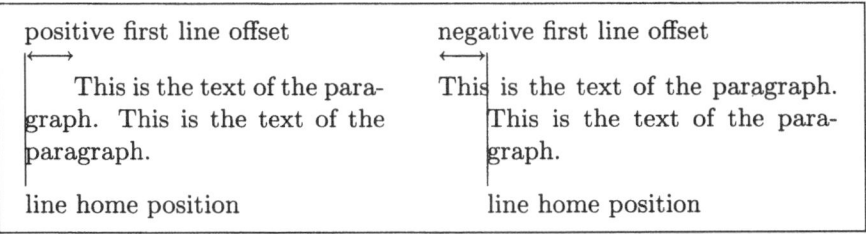

Fig. 51: First line offset for text lines

The line home position is indicated by the vertical lines in the figure. A positive first line offset is often used to emphasize the beginning of a new paragraph graphically, for instance, as in this book. A negative first line offset is often used when formatting tabular material.

The concept of itemization allows for the imaging of an *item identifier* on the first line of a piece of text in positions which are not directly controlled by the line home position or the first line offset. The item identifier, i.e., the character string which shall be graphically separated from the rest of the text, can be positioned with several methods: an *identifier start offset* and an *identifier end offset* can be specified which define the range where the item identifier shall appear. Within this range

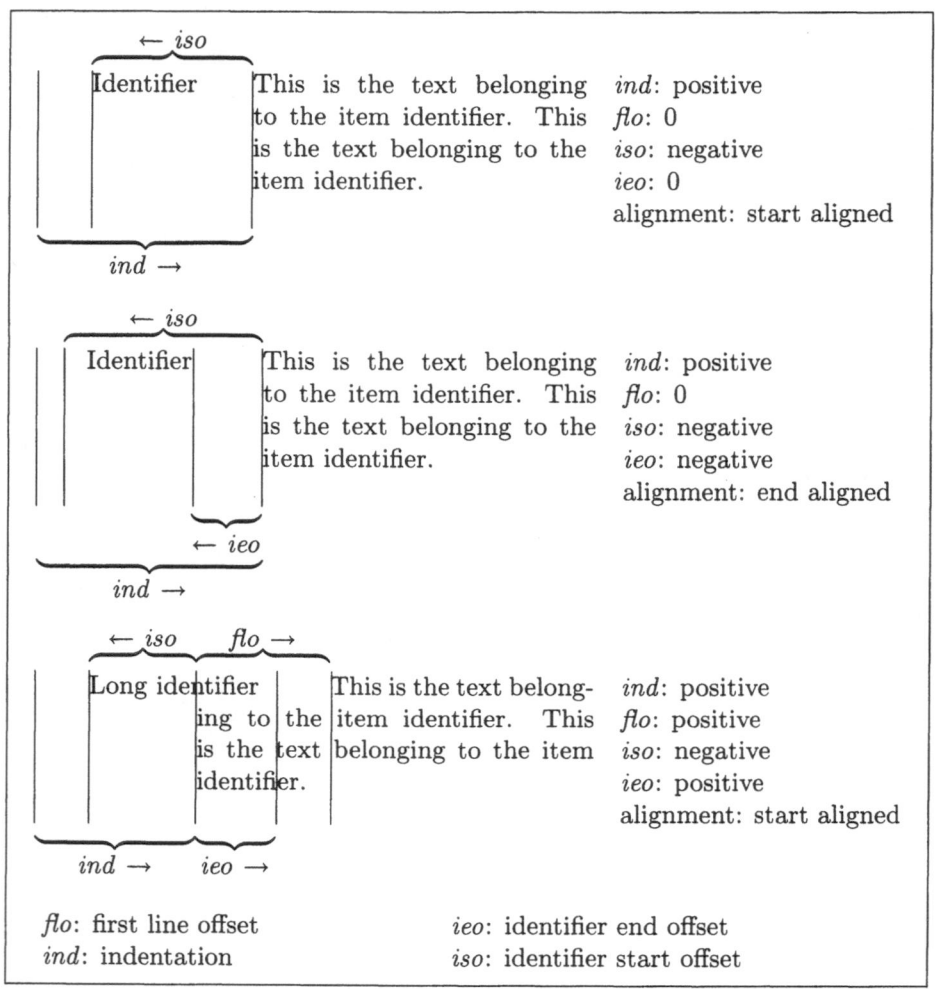

Fig. 52: Examples for itemization

the identifier can be positioned *start aligned* or *end aligned*. Several examples are shown in Fig. 52.

The left parts in this figure show each an item identifier with an associated piece of text, the right parts list the values which where chosen for indentation (*ind*), first line offset (*flo*), identifier start offset (*iso*) identifier end offset (*ieo*) and the identifier alignment method.

The left-most vertical lines indicate the start edge of the positioning area and the next lines to the right specify the (negative) identifier start offset. The right-most lines in the first two examples and the third line in the third example show the line home position. The third line in the second example and the fourth line in the third example indicate the identifier end offset. The last line in the third example indicates the first line offset. The item identifiers in the first and third example are start aligned, the identifier in the second example is end aligned.

Character ordering

The characters of the text in an interchanged ODA document are always arranged in their reading order depending on the language. In an Arabic text which is read from right to left, for instance, the character to the left of a particular character appears *after* that character since Arabic is, of course, both written and read from right to left.

Several languages, however, change their presentation order locally. In Arabic or Hebrew, for example, the "normal" text is written from right to left whereas numbers are (locally) written from left to right, though this change in the writing direction is not performed in the storage order for numbers in an interchanged document.

The ODA Standard therefore provides a way of locally reversing the presentation order by means of the control function SRS which is described on p. 250. This may also be used, for instance, to present a piece of Arabic or Hebrew text embedded in an English text.

Parallel annotation

Two different kinds of script are used in Japan: Kanji which is an ideographic script where a symbol gives no hint as to its pronunciation, and Hiragana and Katakana, which are based on the phonetics of the words. A particular Kanji symbol often has many different meanings and its interpretation depends on the context in which it is used. Similarly, a certain sequence of Kana characters with a particular phonetic value may have different meanings depending on its context.

Though this feature is also found in other languages – the English word "organ", for instance, may denote a part of a human body or a musical instrument – it seems to be a much greater problem in Japanese. Therefore, it is quite common in Japanese texts for Kanji symbols to be annotated with Kana characters to indicate their meaning, if it is not obvious from the context. This annotation scheme is called Ruby.

To facilitate this feature of Japanese scripts, the ODA Standard includes so-called *parallel annotation* which can be performed by means of the control function PTX (see p. 248).

6.1.3   Arrangement of Lines

Within the positioning area, the individual line boxes are assembled in the direction of the line progression. The line home position of the first line box is called the *initial point* of the positioning area.

The distance between the reference lines of two consecutive line boxes is called the *line spacing* (see Fig. 47) and depends on whether a *constant line spacing* or *proportional line spacing* is used.

A constant line spacing is specified by the attribute line spacing (see p. 234) or the control functions SLS and SVS (see p. 246). A proportional line spacing is indicated by the attribute proportional line spacing (see p. 234). When proportional line spacing is performed, the distance between the reference lines of consecutive line boxes is determined by an implementation-dependent algorithm which is not defined in the ODA Standard.

6.1.4   Emphasis of Text

Portions of text may be visually emphasized in character imaging. The following six methods of emphasis are provided:

weight	The contrast or intensity of characters can be modified. Besides the normal weight, *faint* and *bold* intensities are possible.
posture	The posture of characters may be changed between *non-italicized* and *italicized*.
underlining	Characters may be *not underlined*, *underlined* or *doubly underlined*.
blinking	A character image may be steady (not blinking), *slowly blinking* or *rapidly blinking*. Of course, this method of emphasis can only be used on a computer screen.
image inversion	Foreground and background may be interchanged, i.e., characters may have a *positive image* or *negative image*.
crossing out	To mark the deletion of characters in a text, the characters may be *crossed-out*.

The different kinds of emphasis are not described very precisely in the Standard; for instance, it is unspecified what angle of slant italicized characters should have. Particular kinds of emphasis, or combinations thereof, are usually selected by means of the control function SGR (see p. 248).

6.2   Attributes for Character Content

The value of the attribute content architecture class which can be specified
for basic objects or basic object classes or for which a default value will
be derived, indicates whether the content portions associated with the
object or object class contains character content, raster graphics content
or geometric graphics content.

In case of character content, the value of the attribute content architecture class must be an ASN.1 object identifier with the value ⟦2 8 2 6 0⟧
for formatted character content, ⟦2 8 2 6 1⟧ for processable character
content or ⟦2 8 2 6 2⟧ for formatted processable character content. Of
course, the value of the attribute content information for the content portions associated with the basic object or object class must be of the
corresponding content architecture class.

The processing of content portions containing text, in particular the
interpretation of the value of the attribute content information, is specified
by the attributes shown in Fig. 53.

The figure shows which attributes are permitted for a particular content architecture class and whether they are defaultable (d) or non-
mandatory (n). The attributes above the dotted line are the *presentation
attributes* for text. The attributes below the line were already introduced
in Part 2 of the Standard, but their values depend on the content architecture class and are therefore defined in Part 6 as far as character
content is concerned.

6.2.1   Presentation Attributes for Character Content

The attributes below the dotted line in Fig. 53 are presentation attributes
for character content, i.e., they appear as values of the attribute presentation attributes (see p. 101).

They can be specified for basic logical object classes, basic layout object classes, basic logical objects, basic layout objects and presentation
styles which are referenced by these objects or object classes. The attributes represent information how the content portions associated with
these objects or object classes are to be processed by the layout process
and imaging process. Within this section, all presentation attributes for
character content will be described.

| | Content architecture class: | | |
	processable	formatted	formatted processable
alignment	d	d	d
character fonts	n	n	n
character orientation	d	d	d
character path	d	d	d
character spacing	d	d	d
code extension announcers	d	d	d
first line offset	d	d	d
formatting indicator	–	d	d
graphic character sets	d	d	d
graphic character subrepertoire	d	d	d
graphic rendition	d	d	d
indentation	d	–	d
initial offset	–	d	d
itemization	d	d	d
kerning offset	d	d	d
line layout table	d	d	d
line progression	d	d	d
line spacing	d	d	d
orphan size	d	–	d
pairwise kerning	d	d	d
proportional line spacing	d	–	d
widow size	d	–	d
content architecture class	d*	d*	d*
content information	n*	n*	n*
type of coding	d*	d*	d*

*: See also the descriptions of these attributes in Chap. 3

Fig. 53: Attributes of the different content architecture classes for character content

Attributes for the specification and selection of character sets

Several attributes specify and select character sets which are used in the value of the attribute content information for content portions representing character content. These attributes refer to the ISO Standards 2022 (*ISO 7-bit and 8-bit coded character sets – Code extension techniques*), 2375 (*Procedure for registration of escape sequences*), 6937 (*Coded character sets for text communication*) and 7350 (*Registration of graphic character subrepertoires*) which shall not described in detail in this book. For a complete understanding of the attributes code extension announcers, graphic character sets, graphic character subrepertoire and character fonts, the reader is requested to study these Standards.

The value of the attribute code extension announcers consists of a sequence of escape sequences according to ISO 2022:

$$\text{code extension announcers} = \big[\,[\,'ISO\ 2022\ escape\ sequences\,']^+\,\big]$$

This attribute announces the use of code extension features in the text. If it is missing, the sequence of escape sequences announcing the use of the G0 and G2 sets is assumed as its default value.

The value of the attribute graphic character sets consists of a sequence of escape sequences according to ISO 2022 and the register of ISO 2375:

$$\text{graphic character sets} = \big[\,[\,'ISO\ 2022\ escape\ sequences\,']^+\,\big]$$

The attribute specifies one or more graphic character sets which are designated and invoked at the beginning of the text. Other character sets can be designated or invoked within the text by those escape sequences which have been announced by means of the attribute code extension announcers. If the attribute is missing, those escape sequences and shift functions designating and invoking the primary character set of ISO 6937, Part 2, as the G0 set and the supplementary character set of ISO 6937, Part 2, as the G2 set are assumed as its default value.

The value of the attribute graphic character subrepertoire is either 0 or the identifier of a subrepertoire assigned in the register of ISO 7350:

graphic character subrepertoire =
    $(0\,|\,'ISO\ 7350\ subrepertoire\ identifier\,')$

This attribute identifies the subrepertoire of the graphic character repertoire of ISO 6937 used at the beginning of the text; it is only applicable, if the graphic character sets of ISO 6937 are used. Other subrepertoires can be invoked within the text by means of the control function IGS (see p. 248). If the attribute is missing, a default value of 0 is assumed, denoting the full repertoire of the character sets that are designated at the beginning of the text.

The value of the attribute character fonts consists of up to ten parameters with the names primary font, first alternative font, second alternative font, third alternative font, fourth alternative font, fifth alternative font, sixth alternative font, seventh alternative font, eighth alternative font and ninth alternative font. Each parameter has the two sub-parameters font size and font identifier whose values are positive integers:

```
character fonts =
   {[primary font = {font size = 'positive integer'
                     font identifier = 'positive integer'}]
    [first alternative font = {font size = 'positive integer'
                               font identifier = 'positive integer'}]
    [second alternative font = {font size = 'positive integer'
                                font identifier = 'positive integer'}]
    [third alternative font = {font size = 'positive integer'
                               font identifier = 'positive integer'}]
    [fourth alternative font = {font size = 'positive integer'
                                font identifier = 'positive integer'}]
    [fifth alternative font = {font size = 'positive integer'
                               font identifier = 'positive integer'}]
    [sixth alternative font = {font size = 'positive integer'
                               font identifier = 'positive integer'}]
    [seventh alternative font = {font size = 'positive integer'
                                 font identifier = 'positive integer'}]
    [eighth alternative font = {font size = 'positive integer'
                                font identifier = 'positive integer'}]
    [ninth alternative font = {font size = 'positive integer'
                               font identifier = 'positive integer'}]}
```

This attribute designates up to ten fonts which can be used within the text and which are called the primary font, first alternative font, second alternative font, etc. They are selected by means of the attribute graphic rendition (see p. 238) or within the text by means of the control function SGR (see p. 248). The font size for each font is specified in scaled measurement units.

The integer which is used as the value for the sub-parameters font identifier is equal to one of the integers which appears in the attribute fonts list in the document profile (see p. 205). In other words, instead of the actual name of a particular font, a symbolic name – an integer – is used to refer to a font. The mapping of the symbolic name to the actual name is performed by the document profile attribute fonts list.

Specification of the character orientation, character path and line progression

The orientation of the characters and the imaging order of consecutive characters and line boxes is defined by the attributes character orientation, character path and line progression.

The value of the attribute character orientation is either 0°, 90°, 180° or 270°:

character orientation = (0° | 90° | 180° | 270°)

The default value for the character orientation (see also Fig. 45) is 0° which is the intended viewing orientation for the characters. The character orientation applies to all text associated with the object or object class for which the attribute is specified. In particular, the character orientation cannot be changed within the text.

The value of the attribute character path is either 0°, 90°, 180° or 270°:

character path = (0° | 90° | 180° | 270°)

This attribute defines the direction in which consecutive characters are arranged during formatting. Though the character path applies to all text associated with the object or object class for which the attribute is specified, it can be locally reversed by means of the control function SRS as described in the subsection "Character ordering" on p. 225. If the attribute is missing, a default value of 0° is assumed which corresponds to a writing direction from left to right as in Latin scripts.

The value of the attribute line progression is either 90° or 270°:

line progression = (90° | 270°)

This attribute defines the direction in which consecutive line boxes are arranged during formatting. If the attribute is missing, a default value of 270° is assumed, i.e., line boxes are arranged from top to bottom as in Latin scripts. The line progression applies to all text associated with the object or object class for which the attribute is specified. In particular, the line progression cannot be changed within the text.

Definition of the positioning area

The positioning area and the subarea within the positioning area where the text may be rendered is defined by the attributes initial offset, kerning offset and indentation.

The attribute initial offset has the parameters horizontal coordinate and vertical coordinate whose value is a non-negative integer:

initial offset =
 {horizontal coordinate = '*non-negative integer*'
 vertical coordinate = '*non-negative integer*'}

This attribute specifies the distance of the initial point, i.e., of the line home position of the first line box, from the upper left corner of the basic layout object to which the positioning area belongs. The distances are measured in scaled measurement units. If the attribute is missing, the default value is determined as shown in Fig. 54, depending on the direction of the character path and line progression.

character path	line progression	horizontal coordinate	vertical coordinate
0°	90°	0	$v_{blay} - bwex$
	270°	0	$bwex$
90°	90°	$h_{blay} - bwex$	v_{blay}
	270°	$bwex$	v_{blay}
180°	90°	h_{blay}	$bwex$
	270°	h_{blay}	$v_{blay} - bwex$
270°	90°	$bwex$	0
	270°	$h_{blay} - bwex$	0

$bwex$: size of the backward extent of the first line box
h_{blay}: horizontal size of the basic layout object
v_{blay}: vertical size of the basic layout object

Fig. 54: Default value of the attribute initial offset depending on the character path and line progression

For Latin scripts, i.e., for a character path of 0° and a line progression of 270°, the default value for the initial point is such that the first character of a line of text is positioned at the start edge and the first line of text is positioned at the top edge of the area of the basic layout object (see Fig. 46).

The attribute cannot be specified for processable character content, since for this kind of content the formatting process has not yet been carried out. In particular, the text has not yet been broken into line boxes and, therefore, the position of the initial point is still unknown.

The attribute **kerning offset** has the parameters **start edge offset** and **end edge offset** whose values are non-negative integers:

kerning offset =
    {start edge offset = '*non-negative integer*'
     end edge offset = '*non-negative integer*'}

The parameter **start edge offset** defines the distance of the start edge and the parameter **end edge offset** specifies the distance of the end edge of the positioning area to the edge of the area of the basic layout object (see Fig. 46). The distances are measured in scaled measurement units. If the attribute is missing, the default value for both parameters is 0.

The value of the attribute **indentation** is a non-negative integer:

indentation = '*non-negative integer*'

This attribute specifies the distance of the line home position of the line boxes from the start edge of the positioning area (see Fig. 47); the distance is measured in scaled measurement units. A positive value indicates that the complete text shall be indented by that amount during formatting. If the attribute is missing, a default value of 0 is assumed.

The attribute can be considered a layout directive for the formatting process. It therefore cannot be specified for formatted character content for which a layout process has already been carried out.

Specification of character spacing and line spacing

The distances of the characters and lines of a text are specified by the attributes **character spacing, pairwise kerning, line spacing** and **proportional line spacing**.

The value of the attribute **character spacing** is a positive integer:

character spacing = '*positive integer*'

This attribute defines the character spacing which applies at the beginning of the text associated with the object or object class for which the attribute is specified. The character spacing is measured in scaled measurement units; its default value is the equivalent of 120 basic measurement units. The value of this attribute is only applied if a font with constant spacing is used; it is ignored for variable spacing fonts. The character spacing can be changed within the text by means of the control functions SCS and SHS (see p. 244).

The value of the attribute pairwise kerning is either yes or no:

pairwise kerning = (yes | no)

This attribute specifies whether or not pairwise kerning should be applied (see p. 222). The default value is no. The specification concerning pairwise kerning cannot be altered within the text.

The value of the attribute line spacing is a positive integer:

line spacing = 'positive integer'

This attribute defines the line spacing, i.e., the distance of the reference lines of consecutive line boxes, which applies at the beginning of the text associated with the object or object class for which the attribute is specified. The line spacing is measured in scaled measurement units; its default value is the equivalent of 200 basic measurement units. The value of this attribute is only used, if the value of the attribute proportional line spacing is false. The line spacing can be changed within the text by means of the control functions SLS and SVS (see p. 246).

The value of the attribute proportional line spacing is either yes or no:

proportional line spacing = (yes | no)

This attribute indicates whether the line spacing is controlled by the attribute line spacing or the control functions SLS and SVS – if its value is no – or whether the distances between the reference lines of consecutive line boxes shall be determined by the layout process depending on the content in the line boxes (see Sect. 6.1.3) – if the value is yes. If the attribute is missing, a default value of no is assumed.

The attribute can be considered a layout directive for the formatting process. Therefore, it cannot be specified for formatted character content for which a layout process has already been carried out.

Justification

The method used for the justification of a text during formatting is specified by the attribute alignment and, furthermore, by the attribute first line offset which is applied to the first line of a piece of text.

The value of the attribute alignment is either centred, end-aligned, justified or start-aligned:

alignment = (centred | end-aligned | justified | start-aligned)

The meaning of the possible attribute values is described in the subsection "Alignment within a line box" on p. 220. The method applies to the whole text associated with the object or object class for which the attribute is specified. If the attribute is missing, a default value of start-aligned is assumed.

If the value is justified, this alignment method can be locally suppressed by means of the control function JFY (see p. 247). Furthermore, the use of tabulation stops takes precedence over the specified alignment method.

The value of the attribute first line offset is an integer:

first line offset = *integer*

The integer defines the offset of the position point of the first character in the first line box from the line home position (see Fig. 51). A positive value indicates an offset in the direction of the character path, a negative value an offset in the opposite direction. The distance is measured in scaled measurement units.

If the attribute is missing, a default value of 0 is assumed. The first line offset cannot be altered within the text.

Itemization and tabulation

Two special cases for the formatting of text – itemized lists and tabular layout – are controlled by the attributes itemization and line layout table.

The attribute itemization has the parameters identifier alignment, identifier start offset and identifier end offset. The value of the parameter identifier alignment is either no itemization, start-aligned or end-aligned.

The values of the parameters identifier start offset and identifier end offset are numbers:

itemization =
    {identifier alignment = (no itemization | start-aligned | end-aligned)
    identifier start offset = *integer*
    identifier end offset = *integer*}

The attribute indicates whether the start of the text which belongs to the object or object class for which the attribute is specified, shall be formatted as an item identifier and, if so, how the item identifier shall be placed (see p. 223).

If the value of the parameter identifier alignment is no itemization, the text does not contain an item identifier; otherwise it does and the identifier shall either be positioned start aligned or end aligned as shown in Fig. 52.

The parameters identifier start offset and identifier end offset specify offsets from the line home position and define a portion of the line box within which the item identifier is positioned. The offsets are measured in scaled measurement units. A positive value indicates an offset in the direction of the character path, a negative value an offset in the opposite direction (see Fig. 52). The values of these two parameters and the placing of the location of the line home position – specified by the attribute indentation – have to be chosen in such a way that the item identifier can be completely imaged within the positioning area.

The identifier consists of all graphic characters from the start of the text to the first occurrence of the control function CR.

If the attribute is missing, the default value for the parameter identifier alignment is no itemization; the default values for the parameters identifier start offset and identifier end offset are 0, i.e., it is assumed that no item identifier is present.

The value of the attribute line layout table is a set of entries where each entry consists of the parameters tab reference, tab position, alignment and alignment string. The value of the parameter tab reference is a character sequence of up to four decimal digits, the value of the parameter tab position is a non-negative integer and the value of the parameters alignment is either start-aligned, end-aligned, centred or aligned-around. The value of the parameter alignment string is sequence of graphic characters from the character set defined by the attributes graphic character sets and graphic character subrepertoire:

line layout table =
  {[tab reference = '*decimal digits string*'
    tab position = '*non-negative integer*'
    alignment = (start-aligned | end-aligned | centred | aligned-around)
    alignment string = '*character string*']$^+$}

This attribute is used to define tabulation stops along the reference line and to indicate how the text is to be aligned at the tabulation stops (see p. 221). The parameter tab position specifies the distance of the tabulation stop from the start edge of the positioning area measured in scaled measurement units. The parameter alignment indicates the method of alignment which applies at the tabulation stop. If the method is aligned around, the required character string is defined by the parameter alignment string (see Fig. 49).

The parameter tab reference assigns a symbolic name to the tabulation stop. This name is used as the parameter for the control function STAB to identify a particular tabulation stop (see p. 251).

If the attribute is missing, no tabulation stops are defined and, as a consequence, the control function STAB cannot be used in the text associated with the object or object class. A specification or modification of tabulation stops is not possible within the text.

Page break and column breaks

Two attributes – orphan size and widow size – influence page breaks or column breaks in a multicolumn layout. These attributes only apply if a page break or column break would occur when formatting the text associated with the object or object class for which the attribute is specified, i.e., if the text would be split into two or more different blocks or basic pages (see Sect. 6.4.2). The value of the two attributes is a positive integer:

orphan size = '*positive integer*'

widow size = '*positive integer*'

The value of the attribute orphan size defines the minimum number of lines which shall be imaged in the first basic layout object and the value of the attribute widow size defines the minimum number of lines which shall be imaged in the last basic layout object created during the layout process. The default value for the attributes is 1.

The names of the two attributes can be explained as follows: in typography, the term "orphan line" denotes a single line of a paragraph at the end of a page and the term "widow line" a single line of a paragraph at the top of a page. Orphan lines and widow lines are considered bad style in typography and should be avoided. In ODA documents, this could be achieved by specifying a value of 2 or more for these two attributes.

The two attributes can be regarded as layout directives for the formatting process. They cannot be specified for formatted character content since for such content the layout process has already been carried out.

Further specifications

There are two presentation attributes for character content left which have not been described above: **graphic rendition** and **formatting indicator**.

The value of the attribute **graphic rendition** is a sequence of one or more non-negative numbers:

graphic rendition $= [\![\, [\,'non\text{-}negative\ integer'\,]^{+}\,]\!]$

This attribute indicates the type of rendition which is used at the beginning of the text associated with the object or object class for which the attribute is specified.   The possible rendition types are described with the control function SGR (see Fig. 55). The integers in the value of the attribute correspond to the parameter values which are permitted for this control function (see also the restrictions on the combination of parameter values on p. 249). The graphic rendition can be changed within the text by means of the control function SGR. If the attribute is missing, a default value of 0 is assumed, i.e., the implementation-dependent default rendition is selected.

The value of the attribute **formatting indicator** is either **yes** or **no**:

formatting indicator $= (\text{yes} \mid \text{no})$

The presence of this attribute indicates that the content layout process has already been carried out for the text associated with the object or object class for which the attribute is specified, i.e., it cannot be specified if the associated text is in processable form. (The name of this attribute is a bit misleading: it does not indicate whether or not the text has been formatted but rather to what extent.)

The value yes specifies that the text contains control functions representing the effects of the use of the attributes alignment, first line offset, itemization and pairwise kerning and of the use of the control function STAB, i.e., the position of each character in a line box is fully defined by appropriate control functions.

The value no indicates that the effect of these attributes and of the control function STAB has not been reflected by the content layout process by the insertion of appropriate control functions. In this case, the content imaging process has to determine the precise positions of the characters in the line boxes.

6.2.2 Other Attributes

Part 6 of the Standard includes the specification of the values of three attributes which were already introduced in Part 2 – content architecture class, content information and type of coding – since these values depend on the content architecture used.

For character content, the value of the attribute content architecture class is an ASN.1 object identifier whose value is either $[\![2\ 8\ 2\ 6\ 0]\!]$, $[\![2\ 8\ 2\ 6\ 1]\!]$ or $[\![2\ 8\ 2\ 6\ 2]\!]$:

content architecture class $= ([\![2\ 8\ 2\ 6\ 0]\!]\,|\,[\![2\ 8\ 2\ 6\ 1]\!]\,|\,[\![2\ 8\ 2\ 6\ 2]\!])$

This attribute is specified for basic objects and basic object classes and defines the content architecture class of the content portions belonging to them. The value $[\![2\ 8\ 2\ 6\ 0]\!]$ indicates that the text is in formatted form, $[\![2\ 8\ 2\ 6\ 1]\!]$ that it is in processable form and $[\![2\ 8\ 2\ 6\ 2]\!]$ that it is in formatted processable form.

The value of the attribute content information which can be specified for content portions represents the actual pieces of text in an ODA document. Its value consists of a sequence of graphic characters, space characters and control functions, possibly with parameters:

content information $=$
   $[\![\ [('graphic\ character'\,|\,'control\ function'\,|\,'space\ character')]^{+}\,]\!]$

It should be noted that the space character has two roles: Firstly, it is not directly a graphic character, though of course a space between words, for instance, is visible. Secondly, it also acts as a kind of control function since it denotes positions in the text where a line break may occur.

The value of the attribute **type of coding** which can be specified for content portions is an ASN.1 object identifier with the value ⟦2 8 3 6 0⟧:

type of coding = ⟦2 8 3 6 0⟧

This attribute is not of great importance for character content because there exist no different encoding methods for text and its value is therefore fixed. However, since this attribute is introduced in Part 2, a value must be assigned to it. Other coding attributes for text do not exist.

6.3   Control Functions

As mentioned already at several places above, a piece of text, i.e., the value of the attribute **content information**, may not only contain graphic characters but also so-called *control functions* which supply additional information on the visual appearance of the text. The control functions permitted in ODA documents are listed in Fig. 55. The encoding of these control functions shall not be described here: it is defined in the ISO Standard 6429 (*ISO 7-bit and 8-bit coded character sets – Additional control functions for character-imaging devices*).

The figure also indicates whether a control function is allowed in formatted character content (f), processable character content (p) or formatted processable character content (f-p).

The control functions **BPH**, **NBH** and **PTX** are not permitted for formatted character content since they represent instructions on how the formatting process is to be carried out.

The control functions **BS**, **HPB**, **HPR**, **JFY**, **SACS**, **SRCS** and **SSW** are not allowed for processable character content but are inserted into the text by the formatting process.

The two control functions **SOS** and **ST** are only permitted for formatted processable character content since they are used for marking character sequences, including control functions, which have been inserted by the formatting process and are to be eliminated from the text before a reformatting takes place.

All control functions will be described in this section in an order that depends on their semantics.

Control function	Content architecture class		
BPH break permitted here	p	f-p	
BS backspace	f	f-p	
CR carriage return	f	p	f-p
GCC graphic character composition	f	p	f-p
HPB character position backward	f	f-p	
HPR character position relative	f	f-p	
IGS identify graphic subrepertoire	f	p	f-p
JFY no justify	f	f-p	
LF line feed	f	p	f-p
NBH no break here	p	f-p	
PLD partial line down	f	p	f-p
PLU partial line up	f	p	f-p
PTX parallel texts	p	f-p	
SACS set additional character separation	f	f-p	
SCS set character spacing	f	p	f-p
SGR select graphic rendition	f	p	f-p
SHS select character spacing	f	p	f-p
SLS set line spacing	f	p	f-p
SOS start of string	f-p		
SRCS set reduced character separation	f	f-p	
SRS start reverse string	f	p	f-p
ST string terminator	f-p		
STAB selective tabulation	f	p	f-p
SUB substitute character	f	p	f-p
SVS select line spacing	f	p	f-p
SSW set space width	f	f-p	
VPB line position backward	f	p	f-p
VPR line position relative	f	p	f-p

Fig. 55: List of the control functions in character content

Moving the active position

The control functions BS, CR, LF, HPB, HPR, PLD, PLU, VPB and VPR can be used to shift the active position (see p. 218).

BS (backspace)

This control function moves the active position opposite to the direction of the character path. The amount of shift is determined by the most recent occurrence of the control functions SHS (select character spacing) or SCS (set character spacing). If these control functions did not appear previously in the text, the amount is the value of the attribute character spacing (see p. 233).

CR (carriage return)

This control function moves the active position to the line home position of that line box in which it occurs. It is not moved in the direction of the line progression. A CR is usually immediately followed by the control function LF (line feed) to move the active position to the line home position of the next line of text. It is also used to reach the line home position after an item identifier.

The control function CR may only appear in the text if the active position is on the reference line of the line box, and not if the active position has been moved from the reference line by means of the control functions PLD, PLU, VPB or VPR (see p. 243ff.). It is also not permitted in a character string enclosed by the control functions SRS with the parameters 0 or 1 (see p. 250) or in a character string enclosed by the control functions PTX with the parameters 1 or 0 (see p. 248).

LF (line feed)

This control function moves the active position in direction of the line progression. The amount of shift is determined by the most recent occurrence of the control functions SLS (select line spacing) or SVS (select line spacing). If these control functions did not appear previously in the text, the amount is the value of the attribute line spacing (see p. 234).

The control function LF can only be used immediately

- after the control function CR (carriage return),
- after another LF or
- at the start of a text associated with a basic layout object or basic layout object class.

HPB (character position backward)

This control function moves the active position opposite to the direction of the character path. The amount of shift is given by the parameter of the control function which is a positive integer indicating the amount in scaled measurement units. If the parameter is missing, the equivalent of 120 basic measurement units is assumed.

HPR (character position relative)

This control function moves the active position in direction of the character path. The amount of shift is given by the parameter of the control function which is a positive integer indicating the amount in scaled measurement units. If the parameter is missing, the equivalent of 120 basic measurement units is assumed.

This control function has a similar effect as one or more space characters. However, the graphic renditions in effect do not apply to the space effected by HPR, for instance, such a space would not be underlined.

PLD (partial line down) and PLU (partial line up)

The control function PLD starts a sequence of subscript characters or terminates a sequence of superscript characters. Conversely, the control function PLU starts a sequence of superscript characters or terminates a sequence of subscript characters (see p. 222).

If the control function PLD is used its effect must be cancelled by the control function PLU before the control function LF occurs and vice versa.

The control functions do not move the position of the lines used to implement the graphic renditions underlined, doubly underlined or crossed-out when such a graphic rendition was in effect before the control functions occurred.

The actual implementation of these control functions is not described in detail in the Standard; for instance, the amount of shift is unspecified. A particular implementation may even not move the active position at all but use a special font with subscripted or superscripted characters.

VPB (line position backward)

This control function moves the active position opposite to the direction
of the line progression. The amount of shift is given by the parameter
of the control function which is a positive integer indicating the amount
in scaled measurement units. If the parameter is missing, the equivalent
of 100 basic measurement units is assumed. The effect of the control
function must be cancelled before the end of the line is reached.

VPR (line position relative)

This control function moves the active position in the direction of the
line progression. The amount of shift is given by the parameter of the
control function which is a positive integer indicating the amount in scaled
measurement units. If the parameter is missing, the equivalent of 100
basic measurement units is assumed. The effect of the control function
must be cancelled before the end of the line is reached.

Modifying the character spacing and line spacing

The control functions SCS, SHS, SACS, SRCS, SSW, SLS and SVS can
occur in the text to modify the character spacing and line spacing which
is initially defined by the attributes character spacing and line spacing.

SCS (set character spacing)

This control function specifies the character spacing which is to be applied to constant spacing fonts in the subsequent text (see p. 219). The
control function has a positive integer as its parameter which indicates
the character spacing in scaled measurement units. If the parameter is
missing, the equivalent of 120 basic measurement units is assumed. The
specified character spacing remains in effect until one of the control functions SCS or SHS occurs in the subsequent text belonging to the basic
object or object class.

SHS (select character spacing)

This control function specifies the character spacing which is to be applied to constant spacing fonts in the subsequent text. The control function has one parameter whose value is an integer between 0 and 4 with the following meaning:

0:   The character spacing is 120 basic measurement units.
1:   The character spacing is 100 basic measurement units.
2:   The character spacing is 80 basic measurement units.
3:   The character spacing is 200 basic measurement units.
4:   The character spacing is 400 basic measurement units.

If the parameter is missing, the value 0 is assumed. The specified character spacing remains in effect until one of the control functions SCS or SHS occurs in the subsequent text belonging to the basic object or object class.

SACS (set additional character separation)

This control function indicates that a positive inter-character space shall be used in the subsequent text (see p. 219). The control function has an non-negative integer as its parameter which indicates the inter-character space in scaled measurement units. If the parameter is missing, the value 0 is assumed. The specified inter-character space remains in effect until one of the control functions SACS or SRCS occurs in the subsequent text or when the end of the line is reached, i.e., a positive inter-character space is only applied locally.

SRCS (set reduced character separation)

This control function indicates that a negative inter-character space shall be used in the subsequent text (see p. 219). The control function has an non-negative integer as its parameter which indicates the negative inter-character space in scaled measurement units. If the parameter is missing, the value 0 is assumed. The specified inter-character space remains in effect until one of the control functions SACS or SRCS occurs in the subsequent text or when the end of the line is reached, i.e., a negative inter-character space is only applied locally.

SSW (set space width)

This control function explicitly specifies the width of the space character which shall be used in the subsequent text. The control function has a positive integer as its parameter which indicates the space width in scaled measurement units. A default value is not defined: the width of a space character is equal to the character spacing for constant spacing fonts and a property of the font for variable spacing fonts. The specified space width remains in effect until another occurrence of the control function SSW or or until the end of the line is reached, i.e., the space width is only explicitly set locally.

SLS (set line spacing)

This control function specifies the line spacing, i.e., the distance between the reference lines of consecutive line boxes, which is to be used in the subsequent text. The control function has a positive integer as its parameter which indicates the line spacing in scaled measurement units. If the parameter is missing, the equivalent of 200 basic measurement units is assumed. The specified line spacing remains in effect until one of the control functions SLS or SVS occurs in the subsequent text belonging to the basic object or object class.

SVS (select line spacing)

This control function specifies the line spacing, i.e., the distance between the reference lines of consecutive line boxes, which is to be used in the subsequent text. The control function has one parameter whose value is an integer between 0 and 4 or 9 with the following meaning:

0: The line spacing is 200 basic measurement units.
1: The line spacing is 300 basic measurement units.
2: The line spacing is 400 basic measurement units.
3: The line spacing is 100 basic measurement units.
4: The line spacing is 150 basic measurement units.
9: The line spacing is 600 basic measurement units.

If the parameter is missing, the value 0 is assumed. The specified line spacing remains in effect until one of the control functions SLS or SVS occurs in the subsequent text belonging to the basic object or object class.

Controlling line breaks and justification

The three control functions **BPH**, **NBH** and **JFY** can be used to control the line breaking and justification process.

BPH (break permitted here)

This control function indicates a point in the text where a line break may occur when the text is formatted. A line break is usually permitted at any space character in the text and, of course, this need not be indicated explicitly by **BPH**. Sometimes, however, it may be necessary to indicate additional places where a line break would normally be considered invalid.

NBH (no break here)

This control function indicates a point in the text where a line break shall not occur when the text is formatted. Of course, only those places have to be indicated by **NBH** where a line break would usually be considered permitted. For instance, at the space character in the name "S. Price" a line break is usually considered bad typography, i.e., the initials of the name should not be positioned at the end of a line. This can be avoided by inserting a **NBH** immediately after the space character.

JFY (no justify)

This control function may appear at the beginning of a line of text to indicate that the line shall not be justified, i.e., it can be used to locally override the effect of the value justified of the attribute alignment. The control function has one parameter with the value 0.

Additional control functions

The nine control functions **GCC**, **IGS**, **PTX**, **SGR**, **SOS**, **ST**, **SRS**, **STAB** and **SUB** cannot be assigned to one of the groups discussed above and shall be described in this section.

GCC (graphic character composition)

This control function can be used to combine two or more graphic characters to one graphic symbol. The control function has a parameter whose value is either 0, 1 or 2 with the following meaning:

0: The following two graphic characters are combine to one symbol.
1: This parameter starts a string of graphic characters which shall be combined.
2: This parameter terminates a string of graphic characters which shall be combined.

If the parameter is missing, the value 0 is assumed.

IGS (identify graphic subrepertoire)

This control function indicates a subrepertoire of the graphic character repertoire of ISO 6937 which shall be used in the subsequent text until a new subrepertoire is specified or the end of the text associated with the basic object or object class is reached. The control function may have a parameter which is the identifier of the subrepertoire in accordance with the registration procedure specified in ISO 7350. If the parameter is missing, the value 0 is assumed which denotes the entire repertoire of the currently designated graphic character sets.

PTX (parallel texts)

This control function is used to indicate those pieces of text for which the Japanese Ruby annotation should be applied (see p. 225). The control function may have a parameter whose value is either 0, 1 or 2. The parameter 0 is used when the end of the two text strings for parallel annotation is reached. The parameter 1 indicates the start of the two text strings and the parameter 2 delimits the first string from the second. If the parameter is missing, the value 0 is assumed,

SGR (select graphic rendition)

This control function is used to select a particular graphic rendition for the subsequent text. The control function may have one or more param-

Value	Meaning
0	default rendition (implementation dependent)
1	bold
2	faint
3	italicized
4	underlined
5	slowly blinking
6	rapidly blinking
7	negative image
9	crossed-out
10	primary font
11	first alternative font
12	second alternative font
:	
19	nineth alternative font
21	doubly underlined
22	normal density
23	not italicized
24	not underlined
25	steady
26	variable spacing
27	positive image
29	not crossed-out
50	not variable spacing

Fig. 56: Parameters of the control function SGR and their meaning

eters indicating the intended rendition. The meaning of the parameters is shown in Fig. 56 (see also Sect. 6.1.4).

The use of the parameters is cumulative. If, for instance, the control function SGR is first used with the parameter 3 and then with the parameter 4 or if these two parameters are used with the control function, the text will be rendered italicized and underlined. However, the following exceptions hold:

– Parameter values which are mutually exclusive cannot both be in effect, for instance, 4 and 24, 7 and 27 or 26 and 50.
– The value 0 cancels the effects of all previously specified renditions; it cannot be used in combination with any other value.
– If a particular font is selected by one of the parameters 10 to 19, the effects of the possibly specified renditions 1, 2, 3, 22 and 23 are ignored.

The lines created with the parameters 4, 9 and 21 for underlining or crossing-out characters are only raised or lowered if the control function SGR occurs after the control functions PLD or PLU.

SOS (start of string) and ST (string terminator)

The control function SOS starts a string of characters which are to be eliminated before a subsequent layout process is carried out again. The string which may contain control functions besides graphic characters or may even consists of control functions only, is terminated by ST.

This inclusion of character strings within SOS and ST is done for the following reason. During the formatting of a text which was originally in processable form, control functions and maybe characters will be inserted in the text to reflect the result of the formatting process. In particular, the control functions CR (carriage return) and LF (line feed) will be inserted, maybe with a hyphen ("-") if hyphenation takes place, to denote the line breaks effected by the formatting.

On the other hand, the control function CR and LF may have been already present at several places in the text, for instance, where a particular line should terminate in any case. To distinguish such "predefined" line breaks from the ones inserted by the layout process, the latter ones are enclosed by SOS and ST. This allows a later reformatting of the text by removing the so-identified character strings before a new layout process starts.

A line break specified by the control functions LF and CR and enclosed in SOS and ST, is called a *soft line terminator*; if it is not enclosed by SOS and ST, it is called a *hard line terminator*.

Besides the control functions LF and CR, other control functions may be enclosed by SOS and ST, namely all those control functions which can be present both in processable and formatted character content. Only the control functions BS, HPB, HPR, JFY, SACS, SRCS and SSW need not be identified in this way since they are always inserted by the layout process, i.e., their removal from the text before a reformatting is always obvious.

SRS (start reverse string)

This control function is used to indicate a character string whose characters are to be imaged in the direction opposite to that of the immediately preceding text. The control function has a parameter whose value is 1 to denote the start of the string, and 0 to denote its end.

A line break may not occur within the character string. If the control functions PLD, PLU, VPB or VPR appear in the string, their effect must be cancelled before the string terminates. A nesting of the control function SRS is possible, i.e., "...SRS1...SRS1...SRS0...SRS1...SRS0...SRS0" is permitted.

STAB (selective tabulation)

This control function selects a tabulation stop which is defined by means of the attribute line layout table (see p. 236). The control functions has a parameter which is one of the values of the parameter tab reference of this attribute and identifies the tabulation stop. All subsequent text until another occurrence of STAB, or until the end of the line is reached, will be aligned at the tabulation stop as specified by the attribute line layout table.

SUB (substitute character)

This control function is used in the place of a character that has been found invalid or in error.

6.4   The Layout Process for Character Content

The layout process for character content is one of the three content layout processes described in the ODA Standard. It interacts with the document layout process to create the specific layout structure of a document. As for the other layout processes, the Standard does not specify in detail how the layout process for character content has to be carried out, but rather the result of this process is defined.

Slightly simplifying, the layout process for text takes the character content associated with basic logical objects or object classes and creates one or more basic layout objects (blocks or basic pages) where the text is to be imaged. In particular, the dimensions (height and width) of the basic layout objects are determined. They are forwarded to the document layout process which then then determines the precise positions of the basic layout objects within the available area (see p. 170).

The document layout process informs the layout process for character content about the maximum dimensions available for the basic layout object. Within these constraints and on basis of the specifications made by the attributes that apply, the content layout process tries to determine the actual size of the basic layout object. If the text fits into the available area, the content layout process can be successfully carried out.

If the text does not fit in respect of the direction of the character path, the content layout process fails and the document layout process may decide whether a larger area is available to start the content layout process again.

If the text does not fit in respect of the direction of the line progression, the text has to be split into several basic layout objects (see Sect. 6.4.2), i.e., the document layout process has to provide one or more additional areas in which the different portions of the text can be imaged.

It has to be distinguished whether the layout process has to be carried out for processable, formatted processable or formatted character content. A processable or formatted processable text is transformed into formatted or formatted processable form. A formatted text remains in formatted form after the layout process. The text itself, i.e., the value of the attribute content information of the content portions, is usually modified by the layout process (see Sect. 6.4.1).

In general, the following six steps are performed by the layout process for character content:

1. initialization,
2. determination of the initial point of the positioning area,
3. formatting of the content,
4. identification of the content portions,
5. determination of the dimensions of the basic layout object,
6. determination of the value of the attribute initial offset.

Before the central part of the layout process for text – the formatting of the content – is discussed in more detail, the other five steps shall be described briefly.

Initialization

The initialization step is only performed for formatted processable content; it is not required for processable or formatted content. The purpose of this step is the removal of all effects from a previous formatting process. This is done as follows:

- All content portions belonging to the same basic logical object are combined into a single content portion. This is done to avoid unnecessary fragmentation of the content due to a previous content layout process where the content of a basic logical object may have been split into several content portions because several basic layout objects have been created from one basic logical object (see Sect. 6.4.2).
- All character strings starting with SOS and ending with ST are removed from the text, i.e., from the value of the attribute content information of the content portions, since these strings have been inserted from a previous formatting process (see p. 250).
- All control functions BS, HPB, HPR, JFY, SACS, SRCS and SSW are removed from the text since they have been inserted from a previous formatting process.
- The attributes content identifier layout (see p. 89) are removed from the content portions, if present.

After this initialization step the character content is of processable form again, i.e., the layout process is performed as for processable character content. Therefore, it only has to be distinguished between the layout process for processable text and formatted text after initialization.

Determination of the initial point of the positioning area

The initial point of the positioning area (see p. 226) has only to be determined for processable character content; for formatted character content it is specified by the attribute initial offset (see p. 232).

At first, the positions of the start edge and end edge of the area of the basic layout object are determined. Their positions are defined by the values of the attributes character path and line progression. For a writing direction from left to right and top to bottom, the start edge lies at the left-hand side and the top edge at the upper side of this area (see Fig. 46).

Afterwards the start edge and end edge of the positioning area are determined by taking the attribute kerning offset (see p. 233) into account. The distance of the initial point from the start edge of the positioning area is specified by the attribute indentation (see p. 233). It is then known how much space in the direction of the character path is available for the formatting of the text.

Finally, the distance of the initial point from the top edge of the positioning area has to be determined. For this purpose, the size of the backward extent of the first line box must be known. This size, however, depends on the fonts used within the first line of the text and maybe on control functions such as PLU (partial line up) which may influence the size of the backward extent. The precise position of the initial point can therefore only be determined after the text, at least its first line, has been formatted.

Identification of the content portions

After the formatting process for a content portion has been carried out this has to be noted. It is done by adding the attribute content identifier layout with an appropriate value to the content portion. This identifies the content portion as part of the specific layout structure. The value attribute content identifier logical may also be present if the content is in formatted processable form, i.e., if it belongs both to the logical and layout structure.

Determination of the dimensions of the basic layout object

After the formatting of the text as described in Sect. 6.4.1, the minimum size of the area required to render the text is known. This defines the size of the area of the basic layout object where the text is imaged.

For a character path of 0° or 180° the size of the text in this direction corresponds to the horizontal dimension, the size in direction of the line progression the vertical dimension of the basic layout object. For a character path of 90° or 270° (as for Kanji, for example) the size of the text in this direction corresponds to the vertical dimension, the size in direction of the line progression the horizontal dimension of the basic layout object.

If the available area of the basic layout object has been large enough to contain the formatted text, the content layout process has been successful; otherwise, the document layout process should provide a larger available area or the text has to be split into several basic layout objects, i.e., the document layout process should provide additional available areas (see Sect. 6.4.2).

Determination of the value of the attribute initial offset

As the final step of the layout process for character content, the position of the initial point relative to the upper left corner of the basic layout object is explicitly specified by assigning values to the parameters horizontal coordinate and vertical coordinate of the attribute initial offset. Though this attribute need not necessarily be specified – in this case, a default value is determined – it is recommended in the Standard that it always should. Otherwise, the imaging process would become more complicated since this process needs to know the position of the initial point.

6.4.1   Formatting of the Content

The central part of the layout process for character content is the formatting of the text, which shall be described in this section.

The formatting of the content involves in particular

- the positioning of the character images within the line box,
- the determination of line breaks in the case of processable character content, and
- the positioning of the line boxes within the positioning area.

It may be a bit surprising that the ODA Standard applies the term "formatting" also to formatted character content but in this case the formatting process includes only the positioning of the character images within the line box and the positioning of the line boxes within the positioning area; line breaks are not determined. The positioning of characters and lines for formatted character content may be necessary because the receiver of a document does not have the same fonts which were used originally for the formatting of the text. In the following, the formatting process for formatted text shall not be considered but only the more general case of the formatting of processable character content.

It has to be distinguished whether the text is to be in formatted form or in formatted processable form after formatting, i.e., whether or not a receiver of the document shall be able to modify it. If the text shall be transformed to formatted form, the formatting process is easier. In particular, all control functions that are only allowed for processable or formatted processable form are eliminated from the text and control functions such as SOS or ST are not inserted into it.

The formatting of processable content and the generation of formatted processable content is the most general case and will therefore be
discussed in the following. It will be indicated if a difference from the
generation of formatted content exists.

The main effect of the formatting process is the insertion of control
functions into the text, i.e., into the value of the attribute content information, to describe the intended visual appearance of the text. Most of
these control functions will deal with explicit shifts of the active position (see p. 218) to place graphic characters at certain positions on the
positioning area.

Positioning of character images within a line box

The formatting process determines the allocation of the characters to
each line and the dimensions of each line box. However, it may or may
not determine the precise position of each character within a line box. If
not, the precise position of each character is determined by the imaging
process.

At first, we describe the case that each character position is determined
precisely by the formatting process.

The position of the characters depends on the following attributes or
control functions:

– pairwise kerning:
  If the attribute pairwise kerning which applies to the text has the value
  yes, i.e., if the position point of a character is not necessarily placed
  at the escapement point of its preceding character (see p. 222), the
  control functions HPB or HPR are inserted between two characters for
  which a pairwise kerning applies.
– first line offset:
  If the value of this attribute is different from 0, the formatting process
  inserts the control functions HPB and HPR to effect the intended offset
  of the first line of a text (see Fig. 51).
– itemization:
  If the value of the attribute indicates that an item identifier is used
  in the text (see Fig. 52), the identifier is positioned by means of the
  control functions HPB and HPR.
– line layout table and STAB:
  This attribute is used together with the control function STAB to align
  character strings at tabulation stops (see p. 221). The formatting
  process inserts between each STAB and the first graphic character

following it the control functions HPB and HPR to align the character strings according to the specifications.

- alignment:
The effect of this attribute depends on its value. If its value is **start-aligned**, no additional control functions are needed in the text.

If the value is **end-aligned** or **centred**, the formatting process inserts the control function HPR before the first graphic character of each line or after the CR which may terminate an item identifier in the first line.

If the value is **justified**, the justification is effected by the insertion of the control functions SSW, SACS and SRCS with appropriate values. The precise process of justification is implementation-dependent and not described in detail in the Standard. Furthermore, the last line of a portion of text is started with the control function JFY to avoid a justification of this line.

If the precise position of each character in each line box has been determined in this way, the attribute **formatting indicator** receives the value **yes**. If not, i.e., if the determination of the precise positions is deferred to the imaging process, this attribute receives the value **no**.

One reason why the precise positions should not be determined by the formatting process would be that a formatted document is sent to a receiver who does not have the same fonts as the sender. In this case, the precise positions would be of no use to the receiver since he would have to adjust the positions according to the metrics of the characters in his font set anyway. Of course, the font metrics of the sending and receiving system should be "close enough" to provide a reasonable document image after the font substitution.

Line breaks

If the text which is to be formatted requires more space in the direction of the character path than is available within the positioning area, line breaks have to be inserted into the text. This is indicated by the control function CR followed by LF.

In this case, it has to be distinguished whether the result of the layout process is formatted or formatted processable character content. In the latter case, such line breaks are enclosed within the control functions SOS and ST (see p. 250), i.e., the sequence "SOS CR LF ST" is inserted. If hyphenation takes place, the hyphens are also enclosed between SOS and ST.

If the result of the layout process is formatted character content, the line breaks are not enclosed within SOS and ST. However, the control functions BPH and NBH (see p. 247) are eliminated from the text, if present.

The ODA Standard does not specify the line breaking process in detail. As a general rule, however, a line break is permitted at any space character which is not immediately followed by NBH, and at any place where a BPH occurs. Further places may be found by a hyphenation algorithm.

A line break is not allowed as long as subscript or superscript rendition is in effect, within a character string where the character ordering has been reversed by the control function SRS, or within a character string for which the parallel annotation applies.

Parallel annotation

The use of parallel annotation is indicated in the text by the control function PTX (see p. 248). The formatting process inserts the control functions HPB, HPR, VPB and VPR with appropriate parameter values to position the two text strings in parallel to each other.

It has to be distinguished whether the result of the layout process is formatted or formatted processable character content. In the latter case, the control functions PTX are removed from the text. For formatted processable character content, these control functions remain and the new inserted control functions HPB, HPR, VPB and VPR are inclosed within SOS and ST.

Positioning of the line boxes

Consecutive line boxes are positioned in the direction of the line progression. For a constant line spacing, the distance of the reference lines of adjacent line boxes is given by the attribute line spacing; however, it may be changed within the text by the control functions SLS and SVS.

For proportional line spacing, the distance between the reference lines is determined by the formatting process which inserts the control function SLS with appropriate parameter values into the text. The precise method used for proportional line spacing is considered to be implementation-dependent and is not described in the Standard.

6.4.2   Combining and Splitting Content

During the layout process the following situations may arise:

– The content associated with one basic logical object is assigned to one basic layout object. This is the case that was assumed in the preceding section.
– The content associated with one basic logical object is assigned to several basic layout objects. This is the case, for instance, when page breaks occur within the text because the whole text does not fit into the area which was available on the initial page.

  In this case, the text has to be split into two or more content portions. Each basic layout object gets a separate content portion assigned which contains exactly that text which is imaged in the area of the basic layout object.

  The presentation attributes of the second and maybe following basic layout objects receive those values that are valid at the end of the text associated with the preceding basic layout object. If, for example, the attribute line spacing of the first basic layout object has a value of 200 scaled measurement units but the line spacing is changed to 300 scaled measurement units by means of the control function SLS within its associated text, the attribute line spacing for the second basic layout object receives the value 300 scaled measurement units.

  It should also be noted that the values of the attributes orphan size and widow size have to be considered for the second and maybe following basic layout object.
– The content associated with several basic logical objects is assigned to one basic layout object. This is the case if the attribute concatenation with the value concatenated (see p. 123) applies to the basic logical objects following the first one.

  In this case, all presentation attributes specified for the second and maybe following basic logical objects are ignored except the attribute proportional line spacing.
– The content associated with several basic logical objects is assigned to several basic layout objects. This case is a combination of the two preceding cases and is treated accordingly.

It should be noted that these different situations may only arise for character content. For raster graphics or geometric graphics content, the content of one basic logical object is always assigned to one basic layout object.

6.5   The Imaging Process for Character Content

Before the imaging process can be carried out the text has to be transformed to formatted or formatted processable form by the layout process. The task of the imaging process for character content is the transformation of the content portions representing text into a form which is accepted by an imaging device such as a printer or computer screen.

The graphic characters of the text have been assigned to the individual line boxes by the layout process for character content. Whether the precise positions of the character images within the line boxes has been determined also by the layout process depends on the value of the attribute **formatting indicator**. If its value is **yes**, this has been done; if its value is **no**, the imaging process will determine the precise positions. This is done by taking the values of the attributes **pairwise kerning, first line offset, itemization, line layout table** and **alignment** into account as described on p. 256. All characters are positioned relative to the initial point as defined by the attribute **initial offset**.

For the imaging process, the attributes **graphic character sets, character fonts** – with respect to the document profile attribute **fonts list** – and **graphic rendition** and the control functions **IGS** and **SGR** are of importance since they determine the character images which are to be chosen. The control functions **BPH, NBH, PTX, SOS** and **ST** which may be present in formatted processable text are ignored.

The precise implementation of the imaging process depends, of course, on the hardware used for rendition and is therefore not described in the Standard in detail.

7 Part 7: Raster Graphics Content Architectures

This Part of the ODA Standard defines how raster images which may appear in content portions of ODA documents have to be encoded, which attributes relate to such images and how the layout process and imaging process is to be carried out for them.

7.1 The ODA Model for Raster Graphics

A raster image is a two-dimensional array; each element of the array is called a *picture element* or *pel*, for short. Each pel has either the state "set" or "unset". When rendering a raster image on a presentation medium, a particular rectangular area on the medium is associated with each pel and the state of the pel defines the colour of this area.

In the state "unset" the area receives the colour of the background colour of the medium, in the state "set" it receives a different colour. When rendering a raster image on a sheet of paper, the background colour is usually white and the areas marked as "set" will usually be blackened. (It should be noted, that the ODA Standard does not currently support grayscale or multi-coloured pictures.)

The area associated with a pel is called its *reference area*. The heights and widths of the reference areas in a raster image are fixed, but they need not be equal, i.e., a reference area need not be quadratic.

The rendition of a pel depending on its state ("set" or "unset") is not described in detail in the Standard. For instance, it is not specified that the area shall be completely black (in the case of "set") or completely white (in the case of "unset"). This is often impossible due to technical reasons. A laser printer, for example, blackens more or less circular areas which will usually extend beyond the rectangular reference area to avoid "holes" in lines or areas.

Raster images can be encoded in digital form as a sequence of zeros – this may represent the state "unset" – and ones – this may represent the state "set". The sequence of zeros and ones is divided into several substrings where each substring represents a line of the raster image. A raster image of m lines with n pels in each line consists therefore of a sequence of $m \times n$ elements (bits) as shown in Fig. 57.

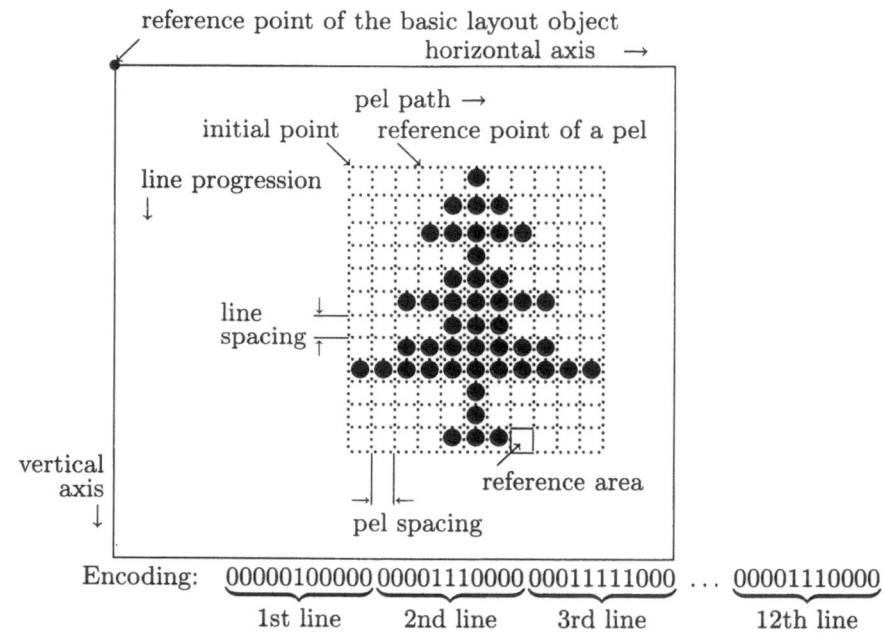

Fig. 57: Example of a raster image

Figure 57 shows a raster images of twelve lines with eleven pels per line and its encoding. Furthermore, several additional terms used with raster images are shown.

The *pel path* defines the direction into which consecutive pels in a line are rendered on the presentation medium. In the example the pel path is from left to right.

The *line progression* defines the direction into which consecutive lines of the image are rendered on the presentation medium. In the example the line progression is from top to bottom.

The *pel spacing* specifies the width of the pels and the *line spacing* the distance of consecutive lines, i.e., these two dimensions define the width and height of the reference area.

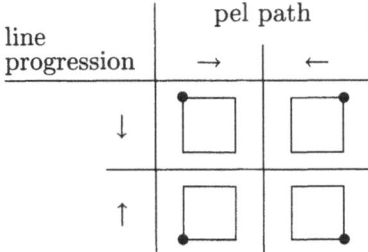

Fig. 58: Position of the reference point (•) in relation to the pel path and line progression

Each reference area has an associated *reference point* which is positioned at one of the four corners of the reference area. Its position depends on the pel path and line progression as shown in Fig. 58.

Figure 57 shows also the edges of the basic layout object to which the raster image belongs, its horizontal and vertical axis and its reference point. This is important since several terms for raster images are defined in relation to the associated basic layout object. For instance, the direction of the pel path is given in relation to the horizontal axis of the basic layout object.

Tiling of raster images

The pel array may be subdivided into a two-dimensional array of non-overlapping rectangular regions called *tiles* as shown in Fig. 59. (The concept of tiled raster graphics is an extension to the initially published Part 7 of ISO 8613. The respective Addendum to Part 7 of the Standard is published in 1991.)

Each tile of a tiled raster image has the same width and height. The width is specified by the number of pels within each tile line and the height by the number of raster lines in each tile. The tiles are ordered in direction of the pel path and line progression as the numbers 1, 2, 3, ... in the Fig. 59 indicate.

Tiles may extend beyond the raster image as is the case for the tiles 1, 2, 3, 6, 9 ... 12 in the example. Such tiles are augmented by the appropriate number of "virtual" pels since all tiles must in principle have the same number of pels. In particular, a so-called *tiling offset* may be specified which defines the distance of the first pel in the first tile to the first pel of the raster image. (In Fig. 59 a positive tiling offset is shown

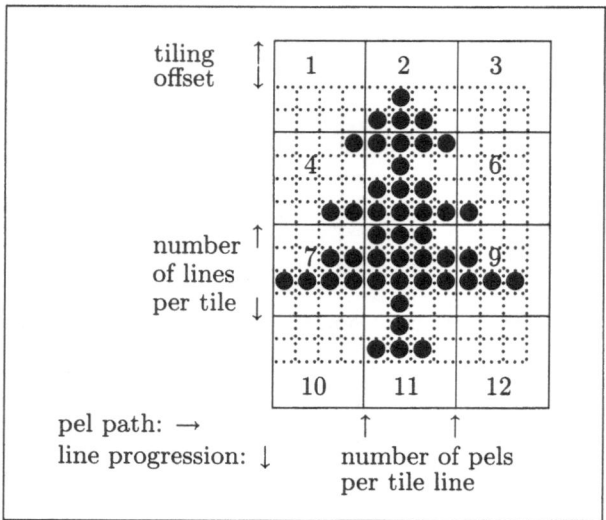

Fig. 59: Tiling of a raster image

in direction of the line progression; however, a positive tiling offset in direction of the pel path is also possible.)

The main purpose of tiling is to provide a more efficient encoding scheme in certain applications: each tile may be encoded according to a different encoding method which may lead to a higher degree of data compression compared to a single encoding scheme applied to the whole raster image. In particular, the tiling of an image has no effect on the layout or imaging process.

7.2   Attributes for Raster Graphics

The value of the attribute content architecture class which can be specified for basic objects or basic object classes or for which a default value will be determined indicates the kind of content (character content, raster graphics content or geometric graphics content) contained in content portions associated with the object or object class.

In case of raster graphics content, the value of this attribute is an ASN.1 object identifier with the value ⟦2 8 2 7 0⟧ (for formatted raster graphics content) or ⟦2 8 2 7 2⟧ (for formatted processable raster graphics content). (The class processable raster graphics content does not exist.) This indicates in particular that the value of the attribute content information for the associated content portions represents a raster image of the indicated class.

How such content portions representing raster images are to be processed, i.e., how the value of the attribute content information has to be interpreted and how the image has to be rendered on a presentation medium is defined by the attributes shown in Fig. 60.

	formatted raster graphics content	formatted processable raster graphics content
coding attributes:		
compression	d	d
number of discarded pels	d	—
number of lines	—	o
number of lines per tile	—	d
number of pels per line	d	m
number of pels per tile line	—	d
tile types	—	d
tiling offset	—	d
presentation attributes:		
clipping	—	d
image dimensions	—	d
initial offset	d	—
line progression	d	d
pel path	d	d
pel spacing	—	d
pel transmission density	d	—
spacing ratio	—	d
Attributes from Part 2:		
content architecture class	d*	d*
content information	o*	o*
type of coding	d*	d*
*: See also the description of this attribute in Chap. 2		

Fig. 60: Attributes of the raster graphics content architectures

It is also indicated in this figure whether an attribute is applicable to formatted raster graphics content, formatted processable raster graphic content or both, whether an attribute is defaultable (d), optional (o) or mandatory (m) or whether it is not applicable at all (—).

The attributes are divided into three groups: the coding attributes which essentially describe how the value of the attribute content information has to be decoded, the presentation attributes which supply the information for the layout process and imaging process, and a third group of attributes which have already been introduced in Part 2 of the Standard but whose values depend on the content architecture.

7.2.1 Coding Attributes for Raster Graphics

Eight attributes – compression, number of discarded pels, number of lines, number of lines per tile, number of pels per line, number of pels per tile line, tile types and tiling offset – are coding attributes for raster graphics content. They can be specified for content portions containing raster graphics and define, together with the attribute type of coding, the encoding of the raster image.

The value of the attribute compression is either compressed or uncompressed:

compression = (compressed | uncompressed)

This attribute is only used if the value of the attribute type of coding specifies that the raster image is encoded according to the CCITT Recommendation T.6 or the two-dimensional encoding of the CCITT Recommendation T.4 (see p. 274). Only these two encodings have a compressed and uncompressed mode. If the attribute is missing, a default value of compressed is assumed.

The value of the attribute number of lines is a positive integer indicating the number of raster lines of the raster image:

number of lines = *positive integer*

This attribute can be specified for formatted processable raster graphics content. It is taken into account during the layout process.

The value of the attribute number of pels per line is a positive integer indicating the number of pels in each line of the raster image:

number of pels per line = *positive integer*

The attribute must be specified for formatted processable raster graphics content. If it is missing for formatted raster graphics content, a default value will be determined. The default value depends on the attribute pel transmission density (see p. 270) as shown in Fig. 61.

Value of pel transmission density	Default value for number of pels per line
1	10368
2	5184
3	3456
4	2592
5	2074
6	1728

Fig. 61: Default value of the attribute number of pels per line in relation to the attribute pel transmission density

These default values are based on CCITT Recommendations for facsimile.

The value of the attribute number of discarded pels is a non-negative integer indicating the number of pels at the beginning of each raster which shall not be rendered:

number of discarded pels = *non-negative integer*

This attribute can only be specified for formatted raster graphics content. If the attribute is missing, a default value of 0 is assumed, if the raster lines can be completely imaged within the area of the basic layout object to which the raster image belongs. If not, i.e., if the raster image is too large in the direction of the raster lines to fit into the area of the basic layout object, half the number of the excess pels is ignored both at the start and at the end of each raster line.

The value of the attribute number of lines per tile is a positive integer:

number of lines per tile = *positive integer*

This attribute can only be specified for formatted processable raster graphics content. If the attribute is missing, a default value of 512 is assumed. The attribute defines the number of raster lines in each tile. It is only applicable if the attribute **type of coding** specifies the tiled encoding for the raster image.

The value of the attribute **number of pels per tile line** is a positive integer:

number of pels per tile line = *positive integer*

This attribute can only be specified for formatted processable raster graphics content. If the attribute is missing, a default value of 512 is assumed. The attribute defines the number of pels within each line of a tile. It is only applicable if the attribute **type of coding** specifies the tiled encoding for the raster image.

The value of the attribute **tile types** is a sequence of elements where each element is either **bitmap encoded, null background, null foreground, T4 one dimensional encoded, T4 two dimensional encoded** or **T6 encoded**:

```
tile type =
   [ (bitmap encoded | null background | null foreground
     | T4 one dimensional encoded | T4 two dimensional encoded
     | T6 encoded) ]
```

This attribute can only be specified for formatted processable raster graphics content. The sequence must contain as many elements as there are tiles in the raster image. Each element defines the encoding scheme for a particular tile according to the order of the tiles (see Fig. 59). If the attribute is missing, the value **T6 encoded** is assumed for all tiles. It is only applicable if the attribute **type of coding** specifies the tiled encoding for the raster image.

The values **null background** or **null foreground** indicate that all pels in the respective tile are "set" or "unset". In this case the tile is not represented in the encoding of the raster image, i.e., in the value of the attribute **content information**. The other values specify that the pels of the tile are encoded in the indicated way.

The value of the attribute **tiling offset** is a sequence of two non-negative integers:

tiling offset = ['*non-negative integer*' '*non-negative integer*']

This attribute can only be specified for formatted processable raster graphics content. If the attribute is missing, a default value of (0,0) is assumed. It is only applicable if the attribute type of coding specifies the tiled encoding for the raster image.

The first number of the sequence specifies the offset of the first pel of the raster image from the first pel of the first tile in direction of the pel path and the second value the offset in direction of the line progression. The offsets are measured in number of pels or number of pel lines, respectively.

7.2.2 Presentation Attributes for Raster Graphics

Eight attributes – clipping, image dimensions, initial offset, line progression, pel path, pel spacing, pel transmission density and spacing ratio – are presentation attributes for raster images, i.e., they may appear with the attribute presentation attributes (see p. 101). They can be specified for basic objects, basic object classes and presentation styles which are referenced by such objects or object classes, and supply information on how the layout process and imaging process for raster images is to be carried out.

The value of the attribute pel path is either 0°, 90°, 180° or 270°:

pel path = (0° | 90° | 180° | 270°)

This attribute defines the pel path (see Fig. 57) relative to the horizontal axis of the basic layout object to which the raster image belongs. The angle is measured counterclockwise. An angle of 0° indicates that the pels in a raster line are processed from left to right parallel to the horizontal axis of the basic layout object. If the attribute is missing, a default value of 0° is assumed.

The value of the attribute line progression is either 90° or 270°:

line progression = (90° | 270°)

This attribute defines the direction in which successive raster lines are to be imaged. The angle is measured counterclockwise relative to the pel path. If the attribute is missing, a default value of 270° is assumed.

The value of the attribute pel spacing is either null, or the attribute has the two parameters length and pel spaces whose value is a positive integer:

pel spacing =
(null | {length = *positive integer*, pel spaces = *positive integer*})

This attribute can be specified for formatted processable raster graphics content and defines the method according to which the distance between consecutive pels shall be determined during the layout process. If the value is null, the so-called *scalable dimensions layout process* is applied, otherwise, the *fixed dimensions layout process* (see Sect. 7.3).

If the value of the parameter length is m and the value of the parameter pel spaces is n, the distance of the reference point of two successive pels is defined as m/n scaled measurement units. If the attribute is missing, a value of 4 is assumed for the parameter length and of 1 for the parameter pel spaces. If one scaled measurement unit corresponds to one basic measurement unit (this is the default value for scaled measurement units), the distance between the reference points of consecutive pels is therefore $\frac{1}{300}$th inch.

The value of the attribute pel transmission density is either 1 BMU (basic measurement unit), 2 BMU, 3 BMU, ... or 6 BMU:

pel transmission density =
(1 BMU | 2 BMU | 3 BMU | 4 BMU | 5 BMU | 6 BMU)

This attribute can be specified for formatted raster graphics content and defines the value for both the pel spacing and line spacing (see Fig 57). If the attribute is missing, a default value of 6 BMU is assumed.

The attribute initial offset has the two parameters horizontal coordinate and vertical coordinate whose values are integers:

initial offset =
{horizontal coordinate = *integer*, vertical coordinate = *integer*}

This attribute can be specified for formatted raster graphics content and defines the offset of the so-called *initial point* – this is the reference point of the first pel in the first raster line as shown in Fig. 57 – from the reference point of the basic layout object. The distances are measured in scaled measurement units. If the attribute is missing, the default value for the initial point depends on the values of attributes pel path and line progression as shown in Fig. 62.

Though negative values for the parameters horizontal coordinate and vertical coordinate are permitted, these should be avoided. Negative val-

| Values of the attributes | | Position of the initial point |
pel path	line progression	(Default value for initial offset)
0°(→) 270°(↓)	270°(↓) 90°(→)	
270°(↓) 180°(←)	270°(←) 90°(↓)	
0°(→) 90°(↑)	90°(↑) 270°(→)	
180°(←) 90°(↑)	270°(↑) 90°(←)	

Fig. 62: Default value for the attribute initial offset in relation to the attributes pel path and line progression

ues are only intended for certain content architectures based on CCITT Recommendations T.73 which provide no other clipping mechanism.

The attribute spacing ratio has the two parameters line spacing value and pel spacing value whose value is a positive integer:

spacing ratio =
{line spacing value = *positive integer*,
  pel spacing value = *positive integer*}

The attribute can be specified for formatted processable raster graphics content. If the value of the parameter line spacing value is m and the value of the parameter pel spacing value is n, the ration m/n defines the ratio between the line spacing and pel spacing (see Fig. 57):

$$\text{line spacing} = \frac{m}{n} \times \text{pel spacing} \qquad .$$

This ratio is observed by the layout process, if the sub-parameter aspect ratio flag of the attribute image dimensions has the value fixed (see p. 272). If the sub-parameter aspect ratio flag has the value variable, the attribute spacing ratio is ignored by the layout process. The default value for the attribute is "line spacing value=1" and "pel spacing value=1".

The value of the attribute clipping consists of two pairs of non-negative integers:

clipping =
$$\big[\, [\!\![\, \text{'non-negative integer'} \quad \text{'non-negative integer'} \,]\!\!]$$
$$[\!\![\, \text{'non-negative integer'} \quad \text{'non-negative integer'} \,]\!\!] \,\big]$$

This attribute determines the subregion of the raster image which is to be taken into account by the layout process and imaging process. For instance, if the first pair has the value $[\!\![\, n_1\ l_1 \,]\!\!]$ and the second pair the value $[\!\![\, n_2\ l_2 \,]\!\!]$, all raster lines preceding line l_1 and following line l_2 are to be ignored and also – within each line – all pels preceding pel n_1 and following pel n_2. Such a selection of a subregion within an image is known as *clipping* in computer graphics. If the attribute is not specified, no clipping takes place, i.e., the complete raster image is considered by the layout process and imaging process.

The value of the attribute image dimensions is either automatic or it is one of the three parameters width controlled, height controlled or area controlled.

The parameter width controlled has the two sub-parameters minimum width and preferred width, the parameter height controlled the two sub-parameters minimum height and preferred height and the parameter area controlled the five sub-parameters minimum width, preferred width, minimum height, preferred height and aspect ratio flag. The values of the sub-parameters minimum width, preferred width, minimum height and preferred height are non-negative integers. The value of the sub-parameter aspect ratio flag is either fixed or variable:

image dimensions =
  (automatic
   | width controlled = {minimum width = *non-negative integer*,
                      preferred width = *non-negative integer*}
   | height controlled = {minimum height = *non-negative integer*,
                      preferred height = *non-negative integer*}
   | area controlled = {minimum width = *non-negative integer*,
                preferred width = *non-negative integer*,
                minimum height = *non-negative integer*,
                preferred height = *non-negative integer*,
                aspect ratio flag = (fixed | variable)})

This attribute specifies the intended dimensions of the basic layout object which is to contain the – maybe clipped – raster image. There are four different methods of specification:

- The parameter **width controlled** defines a minimal and a preferred (maximum) size of the image in direction of the pel path. The value of the sub-parameter **preferred width** must not be less than the value of the sub-parameter **minimum width**.
- The parameter **height controlled** defines a minimal and a preferred (maximum) size of the image in direction of the line progression. The value of the sub-parameter **preferred height** must not be less than the value of the sub-parameter **minimum height**.
- The parameter **area controlled** defines a minimal and a preferred (maximum) size of the image both in direction of the pel path and the line progression. Again, the value of the sub-parameter **preferred width** must not be less than the value of the sub-parameter **minimum width** and the value of the sub-parameter **preferred height** not less than the value of the sub-parameter **minimum height**. The value of the sub-parameter **aspect ratio flag** indicates in addition whether the scaling factor in direction of the pel path is to be the same as in direction of the line progression – if the value is **fixed** – or whether the raster image be be distorted – if the value is **variable**.
- The value **automatic** indicates that the scaling factor shall be the same in both directions. The size of the image in direction of the pel path shall be equal to the dimension of the basic layout object in the respective direction.

The effect of this attribute which can be specified for formatted processable raster graphics content, is described in more detail in Sect. 7.3. If the attribute is missing, a default value of **automatic** is assumed.

7.2.3 Other Attributes for Raster Graphics

Part 7 of the Standard defines also the values of three attributes introduced in Part 2 – **content architecture class**, **content information** and **type of coding** – if the attributes relate to raster graphics content.

The value of the attribute **content architecture class** is an ASN.1 object identifier whose value is ⟦2 8 2 7 0⟧ or ⟦2 8 2 7 2⟧ for raster graphics:

content architecture class = (⟦2 8 2 7 0⟧ | ⟦2 8 2 7 2⟧)

This attribute can be specified for basic objects and basic object classes and indicates the content architecture class of content portions belonging to these objects or object classes. A value of ⟦2 8 2 7 0⟧ specifies that the raster image is in formatted form and a value of ⟦2 8 2 7 2⟧ that it is in formatted processable form.

The value of the attribute type of coding which can be specified for content portions, is an ASN.1 object identifier whose value is either ⟦2 8 3 7 0⟧, ⟦2 8 3 7 1⟧, ⟦2 8 3 7 2⟧, ⟦2 8 3 7 3⟧ or ⟦2 8 3 7 5⟧:

type of coding =
 (⟦2 8 3 7 0⟧ | ⟦2 8 3 7 1⟧ | ⟦2 8 3 7 2⟧ | ⟦2 8 3 7 3⟧ | ⟦2 8 3 7 5⟧)

The value ⟦2 8 3 7 0⟧ indicates that the raster image is encoded according to the CCITT Recommendation T.6, the value ⟦2 8 3 7 1⟧ that it conforms to the one-dimensional encoding of the CCITT Recommendation T.4, the value ⟦2 8 3 7 2⟧ indicates the two-dimensional encoding of CCITT Recommendation T.4, the value ⟦2 8 3 7 3⟧ is used if the so-called *bitmap encoding scheme* applies, and the value ⟦2 8 3 7 5⟧ indicates the tiled encoding. In case of a tiled encoding, each tile of the image may be encoded differently. Which encoding applies to a particular tile is specified by the attribute tile types. The value of the attribute content information is needed in particular to process the value of the attribute content information.

The value of the attribute content information, which can be specified for content portions, represents the actual raster image as an octet string:

content information = '*octet string*'

The encoding used for the octet string is specified by the attribute type of coding and, if required, by the attribute compression. The ODA Standard permits four different encodings: three facsimile encodings according to CCITT Recommendations and the so-called *bitmap encoding scheme*. In this encoding, each pel is represented by one bit in the octet string; the value of the bit is 1 for the state "set" and 0 for the state "unset" (see p. 261). If the number of pels per raster line is not a multiple of eight, an appropriate number of "fill bits" is appended to each line. The attribute number of pels per line is used to recognize such fill bits and to eliminate them before the raster image is processed.

7.3   The Layout Process for Raster Graphics

The layout process for raster graphics is one of the three content layout processes described in the ODA Standard. It interacts with the document layout process to create the specific layout structure of a document. As for the other layout processes, the Standard does not specify in detail how the layout process for raster graphics has to be carried out, but rather the result of this process is defined. In contrast to the layout process for character content, however, an implementation of the algorithm to perform the layout process for raster graphics should be rather straightforward.

The layout process for raster graphics takes the content portion which is associated with a basic logical object or object class and contains a raster image in formatted processable form and creates a basic layout object (block or basic page) where the image is to be displayed. In particular, the dimensions (height and width) of the basic layout object are determined. They are forwarded to the document layout process which then then determines the precise positions of the basic layout object within the available area (see p. 170).

The document layout process informs the layout process for raster graphics about the maximum dimensions available for the basic layout object. Within these constraints and on the basis of the specifications made by the attributes that apply, the content layout process tries to determine the actual size of the basic layout object. If the image fits into the available area, the content layout process can be successfully carried out.

If the raster image does not fit, the content layout process fails and the document layout process may decide whether a larger area is available to start the content layout process again.

There are two different methods for the layout process for raster graphics: the so-called *fixed dimension method* and the *scalable dimension method*. Which of them is to be applied depends on the value of the attribute pel spacing (see p. 269). If its value is null, the scalable dimension methods is used, if the parameters length and pel spaces are given for this attribute, the fixed dimension method. Before these two methods are described in detail, a few terms shall be introduced which are used later on.

h_{aa} shall indicate the horizontal size of the available area (see p. 170) which is given to the content layout process from the document layout process. The size is measured in scaled measurement units.

v_{aa} shall indicate the vertical size of the available area which is given to the content layout process from the document layout process. The size is measured in scaled measurement units.

h_{bl} shall indicate the horizontal size of the basic layout object (block) which is determined by the layout process for raster graphics. The size is measured in scaled measurement units.

v_{bl} shall indicate the vertical size of the basic layout object which is determined by the layout process for raster graphics. The size is measured in scaled measurement units.

n_l shall indicate the number of raster lines in the raster image. If clipping takes place (see p. 271), the clipped-off lines are not included in this number. The number is specified by the attribute **number of lines**. If this attribute is missing, the number is determined when decoding the value of the attribute **content information** which represent the raster image.

n_p shall indicate the number the pels in each raster line. If clipping takes place, the clipped-off pels at the start and end of each line are not included in this number. The number is specified by the attribute **number of pels per line** which must always be specified for formatted raster graphics.

ps shall indicate the pel spacing, i.e., the distance between two adjacent pels in a raster line. The pel spacing is measured in scaled measurement units. The determination of this value is different for the two layout methods.

sr shall indicate the ratio between the distance of the raster lines and the distance of the pels (spacing ratio). This value is specified by the attribute **spacing ratio** (see p. 271) or – if the attribute is missing – by the default value for this attribute.

Therefore, before the layout process for raster graphics starts, the values h_{aa}, v_{aa}, n_l and sr are always known. The value ps depends on the layout method used and the values h_{bl} and v_{bl} are determined as a result of the layout process.

In the following, it is always assumed that the layout process is only concerned with that part of the raster array which remains after clipping, i.e., the clipped-off part of the picture is "invisible" for the layout process.

7.3.1   The Fixed Dimension Method of the Layout Process

If the parameters length and pel spaces are specified for the attribute pel spacing, the fixed dimension method is applied. (It should be noted that the default value for this attribute is length=4 and pel spaces=1, i.e., this method is also used when the attribute is missing.) This layout process is rather simple because the dimensions of the basic layout object can be derived immediately from the values of the attributes pel path, pel spacing, spacing ratio, number of lines and number of pels per line (and the attribute clipping which has been evaluated in advance).

The pel spacing is defined as

$$ps = \frac{\text{Value of the parameter length of the attribute pel spacing}}{\text{Value of the parameter pel spaces of the attribute pel spacing}}$$

which leads, depending on the direction of the pel path, to

Value of pel path 0° or 180°:
$$\begin{cases} h_{bl} = n_p \times ps \\ v_{bl} = n_l \times ps \times sr \end{cases}$$

Value of pel path 90° or 270°:
$$\begin{cases} h_{bl} = n_l \times ps \times sr \\ v_{bl} = n_p \times ps \end{cases}$$

The dimension of the basic layout object in direction of the pel path is the product of the number of pels per raster line and the distance of the pels. The dimension in direction of the line progression is the product of the number of raster lines and the distance of the lines which in turn is given by the product of the distance of the pels and the spacing ratio. If either

$$h_{bl} > h_{aa} \quad \text{or} \quad v_{bl} > v_{aa}$$

holds, the content layout process for the raster image cannot successfully be carried out.

7.3.2   The Scalable Dimension Method of the Layout Process

The scalable dimension method is applied if the attribute pel spacing has the value null. In this case, the pel spacing, i.e., the distance of adjacent pels in a raster line, depends on the value of the attribute image dimensions (see p. 272).

This method uses an additional factor, the so-called *aspect ratio*, defined as

$$ar = \frac{n_p}{n_l \times sr}$$

The aspect ratio indicates the ratio between the dimension of the raster image in direction of the pel path and the dimension in direction of the line progression. The value $as = 2$, for instance, specifies that the picture is twice as wide as high for a pel path of $0°$ and a line progression of $270°$. (The absolute dimensions of the picture are not specified by the aspect ratio.)

The determination of the values h_{bl} and v_{bl} depends furthermore on the value of the attribute image dimensions. According to the four possible parameters of this attribute, the following four procedures can be distinguished:

1. The parameter width controlled is specified for the attribute image dimensions and the sub-parameters minimum width and preferred width indicate the minimum and maximum (preferred) width of the picture.
   The values h_{bl} and v_{bl} are determined in such a way that

   – h_{bl} is chosen as large as possible,
   – the ratio between width and height of the image is maintained and
   – the basic layout object fits into the available area.

   In other words, the maximal h_{bl} is determined which satisfies the following three conditions:

   (1)   minimum width $\leq h_{bl} \leq \min(h_{aa},$ preferred width$)$

   (2) $v_{bl} \leq v_{aa}$

   (3a) $\dfrac{h_{bl}}{v_{bl}} = \dfrac{np}{n_l \times sr}$ or

   (3b) $\dfrac{h_{bl}}{v_{bl}} = \dfrac{n_l \times sr}{n_p}$

   (The function $\min(a, b)$ shall evaluate to the minimum of a and b.) Whether equation (3a) or (3b) is used depends on the direction of the pel path: for a direction of $0°$ or $180°$ equation (3a) holds, for a direction of $90°$ or $270°$ equation (3b). It may, however, be possible that no values h_{bl} and v_{bl} can be found which fulfill these conditions.
   For a better understanding consider the example in Fig. 63.
   The area with the bold border indicates the available area which is given from the document layout process to the content layout process

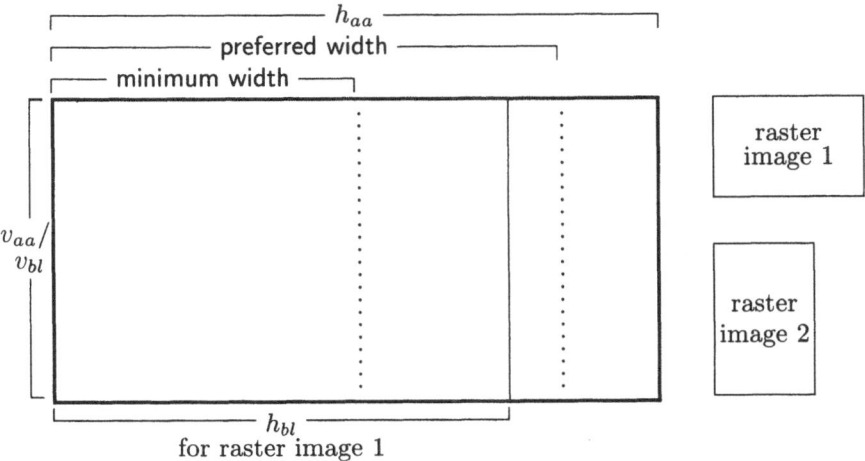

Fig. 63: Scalable dimension method for the parameter **width controlled**

for raster graphics. The ratio between width and height for this area is 2 : 1. The values of the parameters **minimum width** and **preferred width** of the attribute **image dimensions** are shown graphically. The minimum width is a half, the preferred width $\frac{5}{6}$ of the width of the available area. In particular, the minimum width equals the value v_{aa} and the preferred width is $\frac{5}{3}v_{aa}$.

Beneath the available area two raster images are shown where the aspect ratio of the first is 3 : 2 and of the second 2 : 3. (It shall be assumed that the images have a pel path of 0° and a line progression of 270°.)

The situation at the start of the layout process can therefore be summarized as follows:

$$v_{aa} = x \text{ SMU} \quad \text{(the actual value of } x \text{ is not}$$
$$\text{important for the example)}$$

$$h_{aa} = 2x \text{ SMU}$$

$$\text{minimum width} = x \text{ SMU}$$

$$\text{preferred width} = \tfrac{5}{3}x \text{ SMU}$$

$$ar = \tfrac{3}{2} \quad \text{for raster image 1}$$

$$ar = \tfrac{2}{3} \quad \text{for raster image 2}$$

Only the ratio ar, not the actual values of n_p, n_l and sr are important for the example.

For raster image 1, the layout process will compute the value of v_{bl} equal to the height of the available area ($v_{bl} = v_{aa} = x$ SMU) and the value of h_{bl} equal to one and a half times the height ($h_{bl} = \frac{3}{2}v_{bl} = \frac{3}{2}x$ SMU). It can be easily checked that this is the largest value of h_{bl} which satisfies the conditions (1), (2) and (3a).

For raster image 2, the layout process can determine no values of h_{bl} and v_{bl} which satisfy all three conditions since the image is a factor of $\frac{3}{2}$ higher than it is wide, i.e., the value of v_{aa} should be at least this factor larger than the minimum width in order to satisfy condition (2). Therefore, the layout process would fail for this image.

For an aspect ratio greater than $\frac{5}{3}$, the layout process would use the preferred height (h_{bl} = preferred height), whereas the vertical size of the available area would not be needed completely ($v_{bl} < v_{aa}$).

2. The parameter height controlled is specified for the attribute image dimensions and the sub-parameters minimum height and preferred height indicate the minimum and maximum (preferred) height of the picture.

The values h_{bl} and v_{bl} are determined in such a way that

– v_{bl} is chosen as large as possible,
–   the ratio between width and height of the image is maintained and
–   the basic layout object fits into the available area.

In other words, the maximal v_{bl} is determined which satisfies the following three conditions:

(4)   minimum height $\leq v_{bl} \leq \min(v_{aa},$ preferred height$)$

(5) $h_{bl} \leq h_{aa}$

(6a) $\dfrac{h_{bl}}{v_{bl}} = \dfrac{np}{n_l \times sr}$ or

(6b) $\dfrac{h_{bl}}{v_{bl}} = \dfrac{n_l \times sr}{n_p}$

Whether equation (6a) or (6b) is used depends on the direction of the pel path: for a direction of 0° or 180° equation (6a) holds, for a direction of 90° or 270° equation (6b). It may, however, be possible that no values h_{bl} and v_{bl} can be found which fulfill these conditions.

For a better understanding consider the example in Fig. 64.

The area with the bold border indicates the available area which is given from the document layout process to the content layout process for raster graphics. The ratio between width and height for this area is 2 : 3. The values of the parameters minimum height and preferred

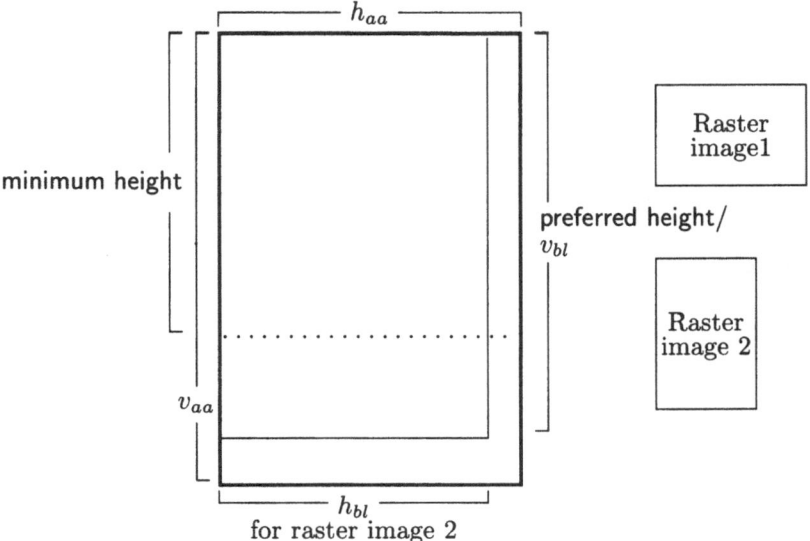

Fig. 64: Scalable dimension method for the parameter **height controlled**

height of the attribute **image dimensions** are shown graphically. The minimum height is $\frac{2}{3}$ and the preferred height $\frac{8}{9}$ of the height of the available area. In particular, the minimum height is equal to the value of h_{aa} and the preferred height is $\frac{4}{3} h_{aa}$.

Beneath the available area, two raster images are shown where the aspect ratio of the first is 3 : 2 and of the second 2 : 3. (It shall be assumed that the images have a pel path of $0°$ and a line progression of $270°$.)

The situation at the start of the layout process can therefore be summarized as follows:

$$v_{aa} = x \text{ SMU} \quad \text{(the actual value of } x \text{ is not}$$
$$\text{important for the example)}$$
$$h_{aa} = \tfrac{2}{3} x \text{ SMU}$$
$$\text{minimum height} = \tfrac{2}{3} x \text{ SMU}$$
$$\text{preferred height} = \tfrac{8}{9} x \text{ SMU}$$
$$ar = \tfrac{3}{2} \quad \text{for raster image 1}$$
$$ar = \tfrac{2}{3} \quad \text{for raster image 2}$$

Only the ratio ar, not the actual values of n_p, n_l and sr are important for the example.

For raster image 2, the layout process will compute the value of v_{bl} equal to the preferred height ($v_{bl} = v_{aa} = \frac{4}{3} h_{aa}$) and the value of h_{bl} equal to $\frac{8}{9}h_{aa}$ since the following relation holds for the aspect ratio:

$$\frac{h_{bl}}{v_{bl}} = \frac{\frac{8}{9}h_{aa}}{\frac{4}{3}h_{aa}} = \frac{2}{3} = ar$$

For raster image 1, the layout process can determine no values of h_{bl} and v_{bl} which satisfy all three conditions (4), (5) and (6a) since the image is a factor of $\frac{3}{2}$ higher than wide, i.e., the value of h_{aa} should be at least this factor greater than the minimum width in order to satisfy condition (5). Therefore, the layout process would fail for this image.

3. The parameter **area controlled** is specified for the attribute **image dimensions** and the sub-parameters **minimum height** and **preferred height** indicate the minimum and maximum (preferred) height of the picture and the sub-parameters **minimum width** and **preferred width** the minimum and maximum (preferred) width of the picture. Furthermore, the sub-parameter **aspect ratio flag** has either the value **fixed** or **variable**. Firstly, the following two conditions have to be satisfied:

(7)   minimum height $\leq v_{aa}$

(8)   minimum width $\leq h_{aa}$

This means that the available area must be large enough to contain the image. The further processing method depends on the value of the sub-parameter **aspect ratio flag**.

If its value is **variable**, a distortion of the aspect ratio of the image is allowed. In this case – as long as conditions (7) and (8) are satisfied – the layout process will always succeed with the values $h_{bl} =$ **preferred width** and $v_{bl} =$ **preferred height**. The picture will be distorted for

$$\text{aspect ratio} \neq \frac{\min(h_{aa}, \textsf{preferred width})}{\min(v_{aa}, \textsf{preferred height})} \quad \text{(pel path } 0° \text{ or } 180°\text{)}$$

or

$$\text{aspect ratio} \neq \frac{\min(v_{aa}, \textsf{preferred height})}{\min(h_{aa}, \textsf{preferred width})} \quad \text{(pel path } 90° \text{ or } 270°\text{)}$$

If the sub-parameter **aspect ratio flag** has the value **fixed**, no distortion of the aspect ratio of the image is permitted. In this case, the layout process determines the greatest possible values for h_{bl} and v_{bl} which satisfy the following conditions:

(9)   minimum width $\leq h_{bl} \leq \min(h_{aa},$ preferred width)

(10)   minimum height $\leq v_{bl} \leq \min(v_{aa},$ preferred height)

(11a) $\dfrac{h_{bl}}{v_{bl}} = \dfrac{np}{n_l \times sr}$ or

(11b) $\dfrac{h_{bl}}{v_{bl}} = \dfrac{n_l \times sr}{n_p}$

Whether equation (11a) or (11b) applies, depends again on the direction of the pel path.

An example for this method shall not be given since this method is essentially a combination of the two methods above and the examples given there should be sufficient to understand the layout process for the parameter **area controlled**.

4.   The parameter **automatic** is specified for the attribute **image dimensions**. In this case, the layout process tries to adjust the width of the image to the width of the available area. The height of the image will be such that the aspect ratio of the picture is observed, i.e., the following conditions apply to this method:

(12) $h_{bl} = h_{aa}$

(13a) $v_{bl} = \dfrac{h_{bl}}{\text{aspect ratio}}$

(13b) $v_{bl} = h_{bl} \times \text{aspect ratio}$

Equation (13a) applies for a pel path of $0°$ or $180°$, otherwise equation (13b).

This method can be considered a special case of the first method when taking there the values

minimum width = preferred width = h_{aa}

It should be noted that for all four methods the values of h_{bl} and v_{bl} must be integral multiples of one scaled measurement unit.

7.4   The Imaging Process for Raster Graphics

The image process for raster graphics is rather straightforward and consists essentially of a marking of certain areas on the rendition medium, for instance, by blackening certain areas on a sheet of paper. This is, of course, dependent on the hardware for the rendition medium and therefore the ODA Standard does not describe this process in further detail.

Two cases have to be distinguished for the imaging process:

1. The raster image is in formatted form.

    In this case, only those pels of the image are rendered which remain when taking the value of the attribute **number of discarded pels** into account. The first remaining pel is positioned at the place specified by the attribute **initial offset** (see Fig. 57).

2. The raster image is in formatted processable form.

    Firstly, those lines and rows of the raster image are removed which are specified by the value of the attribute **clipping**. Then the initial point of the image is determined by evaluating the attributes **pel path** and **line progression** (see Fig. 62), i.e., one of the four corners of the basic layout object will be used as the initial point of the image.

    The pel spacing is determined by dividing the width of the basic layout object by the number of pels per raster line, ignoring those pels in a line which have been clipped-off. The line spacing is determined by dividing the height of the basic layout object by the number of raster lines, again ignoring clipped-off lines.

In both cases, only those pels are imaged on the rendition medium which lie completely within the area of the basic layout object.

8 Part 8: Geometric Graphics Content Architectures

This Part of the ODA Standard defines how geometric graphics which may appear in content portions of ODA documents have to be encoded, which attributes relate to such pictures and how the layout process and imaging process is to be carried out for them.

8.1 The ODA Model for Geometric Graphics

The ODA Standard does not introduce a new model for geometric graphics but the model from the ISO Standard 8632 (*Computer Graphics – Metafile for the Storage and Transfer of Picture Description Information*) is carried over almost completely. This model shall not be described here in detail; the reader is requested to read this Standard or books describing this Standard for additional information. A geometric picture in the ODA sense is simply a picture encoded as a *Computer Graphics Metafile* (CGM) with the following two exceptions from ISO 8632:

- The rules for determining the default values of CGM parameters are different.
- A CGM may only contain one picture.

Furthermore, the following should be noted: In a CGM, so-called *virtual device coordinates* (VDCs) are used. ODA has also a coordinate system but it is different from the CGM system. Therefore, the relation between the two coordinate systems has two be specified.

This is done in the ODA Standard as follows: The so-called *region of interest* in a CGM may be given by $l = (x_1, y_1)$ and $r = (x_2, y_2)$. The region of interest is that part of a picture which is to be imaged on a rendition medium. The region of interest is specified in VDC coordinates. For a better understanding consider the example in Fig. 65.

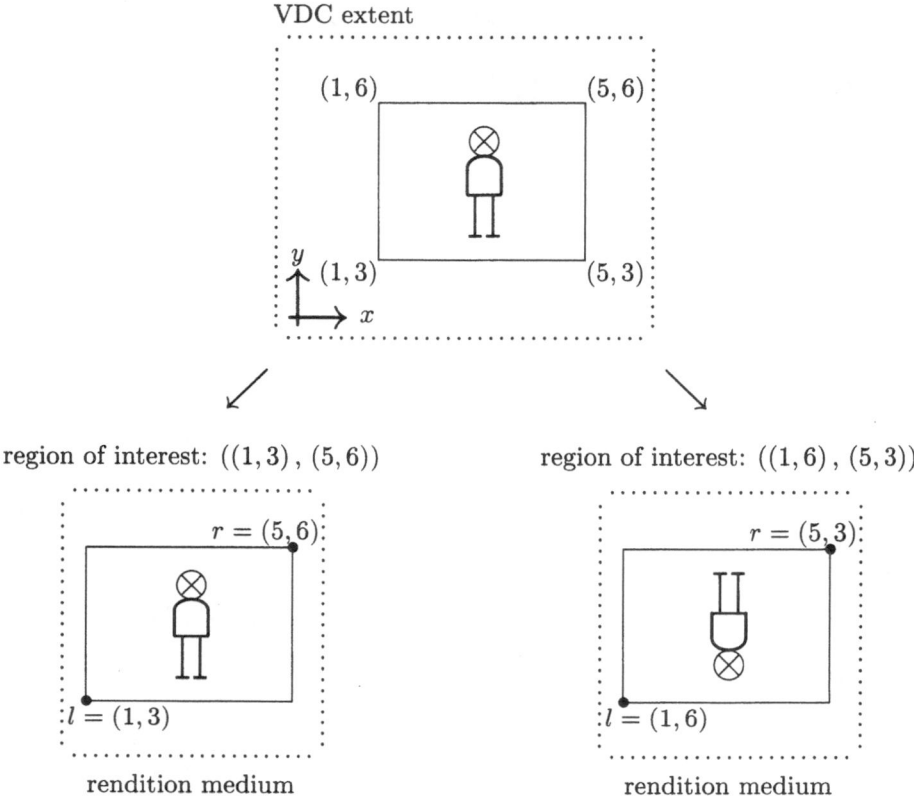

Fig. 65: Mapping of the CGM coordinates onto the rendition medium depending on the specification of region of interest

The picture in the upper half of Fig. 65 may be contained in a given CGM. The x-axis of the virtual device coordinates may extend from left to right ($\rightarrow$) and the y-axis from top to bottom ($\uparrow$). That part of the picture which is to be imaged may be between the x coordinates 1 and 5 and between the y coordinates 3 and 6.

There are two different possibilities to specify the region of interest by two points l and r: either by $l = (1,3)$, $r = (5,6)$ or by $l = (1,6)$, $r = (5,3)$. Figure 66 shows the effect of the two specifications: with the first method the picture will appear as defined in the CGM and with the second method it will appear "upside down".

It should be noted that in the example of Fig. 66 it was assumed that the x-axis of the basic layout object containing the picture is parallel to the x-axis of the CGM, i.e., the x-axis of the basic layout object is horizontal ($\leftarrow$ or $\rightarrow$). Otherwise, the picture would have been rotated.

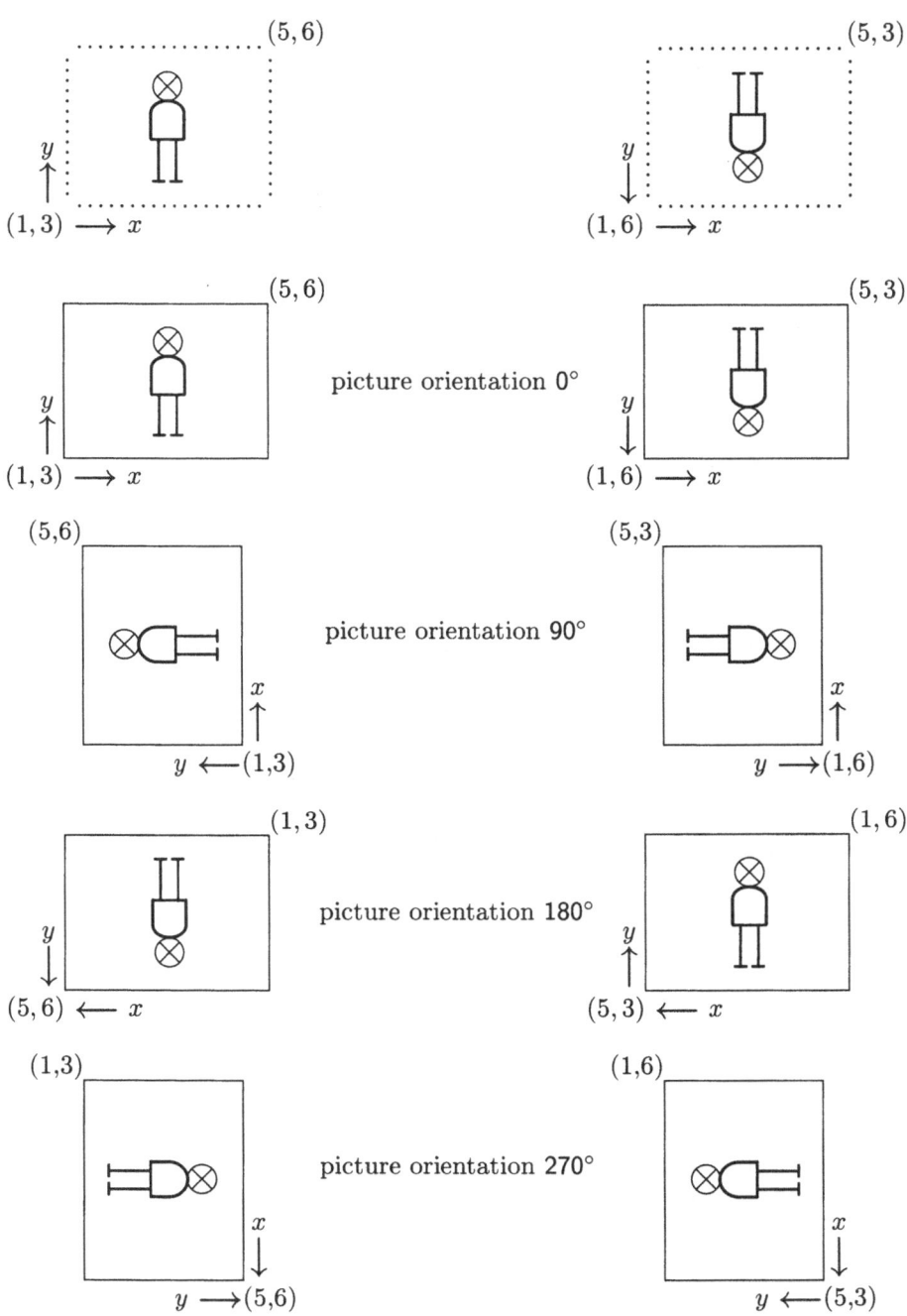

Fig. 66: Rendition of the picture depending on the specification of the region of interest and the picture orientation

In the following, a horizontal orientation of the x-axis of the basic layout object is always assumed. If the x-axis has a vertical orientation, all examples have to be modified accordingly.

By means of the attribute picture orientation (see p. 290) an additional rotation of the picture can be specified. At the top of Fig. 66 the normal orientation of the picture is shown, followed by the four different choices for the picture orientation. The solid borders around the pictures indicate the four edges of the basic layout object containing the picture.

8.2   Attributes for Geometric Graphics

The value of the attribute content architecture class which can be specified for basic objects or basic object classes or for which a default value will be determined indicates the kind of content (character content, raster graphics content or geometric graphics content) contained in content portions associated with the object or object class.

In case of geometric graphics content, the value of this attribute is an ASN.1 object identifier with the value [2 8 2 8 0] indicating formatted processable geometric graphics content. (The classes processable geometric graphics content or formatted geometric graphics content do not exist.) This indicates in particular that the value of the attribute content information for the associated content portions represents a geometric picture in the form of a Computer Graphics Metafile.

How such content portions representing geometric pictures are to be processed, i.e., how the value of the attribute content information has to be interpreted and how the picture has to be rendered on a presentation medium is defined by the attributes shown in Fig. 67. A distinction is made between attributes from Part 2 and presentation attributes for geometric graphics content.

Of the twelve presentation attributes the nine attributes below the dotted line specify values for CGM parameters. (The CGM Standard uses also the term *parameter* but with a slightly different meaning.)

Presentation attributes	Attributes from Part 2
picture dimensions	content architecture class
picture orientation	content information
region of interest specification	type of coding
. . . .	
colour representations	
edge rendition	
filled area rendition	
geometric graphics encoding announcer	
line rendition	
marker rendition	
text rendition	
transformation specification	
transparency specification	

Fig. 67: Attributes of the geometric graphics content architecture

8.2.1   Data Types of Attribute Values for Geometric Graphics

Several data types are used with the values of attributes of geometric graphics content which do not appear in other Parts of the ODA Standard. In particular, these are data types which are defined in the CGM Standard such as "*VDC Pair*" which is used to specify a pair of coordinates in the virtual device coordinate space of the CGM.

As with the general concepts of CGM, these data types shall not be described here. All "imported" data types are written in upper-case letters – at least their first letter will be upper-case – and can therefore be easily identified in the text.

8.2.2   Presentation Attributes for Geometric Graphics

Twelve attributes – picture dimensions, picture orientation, region of interest specification, colour representations, edge rendition, filled area rendition, geometric graphics encoding announcer, line rendition, marker rendition, text rendition, transformation specification and transparency specification – are presentation attributes for geometric graphics content, i.e., they may appear within the attribute presentation attributes (see p. 101). They can be specified for basic logical object classes, basic layout object classes, basic layout objects and basic logical objects and for presentation styles which are referenced by such objects or object classes, and supply information on how the layout process and imaging process for geometric pictures is to be carried out.

Firstly, the three attributes which are not directly related to the CGM shall be described. These are the attributes **picture dimensions**, **picture orientation** and **region of interest specification**.

The value of the attribute **region of interest specification** is either automatic or the parameter **rectangle**. The value of this parameter consists of two pairs of virtual device coordinates where the first pair defines the left-side corner and the second pair the right-side corner of the region of interest (see Fig. 65):

region of interest specification =
    (**automatic** | **rectangle** = ⟦ ' *VDC Pair* '† ' *VDC Pair* '† ⟧)

Each of the coordinate pairs is represented as an octet string according to the so-called *binary encoding scheme* of ISO 8632, Part 3. In other words, the encoding scheme for the coordinates is the same used in the encoding of the CGM itself and not the usual encoding of integers in ODA documents. Therefore, a dagger (†) has been added to the data type *VDC Pair*. The binary encoding scheme of the CGM Standard is used for many parameter values of the geometric graphics attributes; it will always be indicated by a dagger.

If the value of the attribute is **automatic** – this is also the default value in case the attribute is missing – the region of interest is the coordinate space of the virtual device coordinates, i.e., a selection of a subregion of the picture does not take place.

The attribute **region of interest specification** plays a similar role as the attribute **clipping** for raster graphics content since it specifies also a clipping frame for the picture. The following difference should be noted, however: For raster graphics, all pels outside the area specified by the attribute **clipping** may be ignored completely. For geometric graphics, is is not allowed to ignore all objects whose coordinates are outside the region of interest. For instance, it may be possible that the end points of a line are completely outside this region, but a part of the line may pass through it. This part must, of course, be imaged on the rendition medium.

The value of the attribute **picture orientation** is either 0°, 90°, 180° or 270°:

picture orientation = (0° | 90° | 180° | 270°)

The value of this attribute defines the orientation of the picture as shown in Fig. 66. It also specifies which corner of the basic layout object shall be coincident with the first corner of the region of interest (denoted as "*l*" in Fig. 65): For 0° it is the lower left corner of the basic layout object, for 90° the lower right corner, for 180° the upper right corner and for 270° the lower left corner.

The value of the attribute picture dimensions is either automatic or one of the three parameters width controlled, height controlled or area controlled. The parameter width controlled has the two sub-parameters minimum width and preferred width, the parameter height controlled has the two sub-parameters minimum height and preferred height, and the parameter area controlled has the five sub-parameters minimum height, preferred height, minimum width, preferred width and aspect ratio flag. The values of the sub-parameters minimum height, preferred height, minimum width and preferred width are non-negative integers and the value of the sub-parameters aspect ratio flag is either fixed or variable:

picture dimensions =
   (automatic
    | width controlled = {minimum width = *non-negative integer*,
                         preferred width = *non-negative integer*}
    | height controlled = {minimum height = *non-negative integer*,
                         preferred height = *non-negative integer*}
    | area controlled = {minimum width = *non-negative integer*,
                    preferred width = *non-negative integer*,
                    minimum height = *non-negative integer*,
                    preferred height = *non-negative integer*,
                    aspect ratio flag = (fixed | variable)}})

This attribute specifies the intended size of the picture or, more precisely, of the basic layout object containing the picture according to one of the following four methods:

- The parameter width controlled defines a minimal and a preferred (maximum) width of the basic layout object. The value of the sub-parameter preferred width must not be less than the value of the sub-parameter minimum width.
- The parameter height controlled defines a minimal and a preferred (maximum) height of the basic layout object. The value of the sub-parameter preferred height must not be less than the value of the sub-parameter minimum height.

- The parameter **area controlled** defines a minimal and a preferred (maximum) width and height of the basic layout object. Again, the value of the sub-parameter **preferred width** must not be less than the value of the sub-parameter **minimum width** and the value of the sub-parameter **preferred height** not less than the value of the sub-parameter **minimum height**. The value of the sub-parameter **aspect ratio flag** indicates in addition whether the scaling factor in direction of the x-axis is to be the same as in direction of the y-axis – if the value is fixed – or whether the picture be be distorted – if the value is **variable**.

- The value **automatic** indicates that the scaling factor shall be the same in both directions. The width of the basic layout object shall be equal to the width of the available area.

This attribute will be discussed in more detail when the layout process for geometric graphics is described (see Sect. 8.3). If the attribute is missing, its default value is **automatic**.

Now the remaining presentation attributes shall be described which specify default values for parameters in CGM pictures. The basic idea is as follows: A CGM contains many parameters to which values must be assigned when the picture is to be imaged on a particular presentation medium. For instance, the width of lines in the picture have to be specified, character fonts for text in the picture have to be selected or patterns to fill areas have to be defined. Such specifications depend often on the rendition hardware and are therefore not contained in the CGM itself. The receiver of a CGM will usually assign values to such parameters whenever he or she wants to image the picture.

To enable such an assignment, the ODA Standard provides nine attributes: **colour representations, edge rendition, filled area rendition, geometric graphics encoding announcer, line rendition, marker rendition, text rendition, transformation specification** and **transparency specification**. These attributes usually contain many parameters and sub-parameters where each parameter or sub-parameter corresponds to a CGM parameter.

For example, the attribute **line rendition** contains the parameter **Line Width** which is used to assign a value to the CGM parameter **Line Width**. During the processing of the CGM, the layout process or imaging process will use this value to determine the width of lines.

Although the CGM Standard specifies also default values for the CGM parameters, these values are not used in the ODA Standard but the ODA Standard defines its own default values. (In many cases, however, the ODA default values are the same as in the CGM Standard.)

In the following, the attributes are introduced rather briefly. In particular, the meaning of the parameters is usually not explained, if they correspond to CGM parameters. The meaning of these parameters is defined in the ISO Standard 8632. If a reader is not familiar with the CGM Standard, some of the explanations below may be difficult to understand since several terms from the CGM Standard are used but not defined in the text.

The following notation – as in Part 8 of the ODA Standard – is used: Whenever the name or parts of the name of a parameter start with an upper-case letter (such as the parameter Line Width above), this indicates that this ODA parameter corresponds directly to a CGM parameter with the same name defined in ISO 8632. The same notation is also used for data types as mentioned already in Sect. 8.2.1.

The attribute geometric graphics encoding announcer consists of the eleven parameters Colour Index Precision, Colour Precision, Colour Selection Mode, Colour Value Extent, Index Precision, Integer Precision, Maximum Colour Index, Real Precision, VDC Integer Precision, VDC Real Precision and VDC Type. All parameters are optional:

geometric graphics encoding announcer =
    { [Colour Index Precision = $(8^\dagger \,|\, 16^\dagger \,|\, 24^\dagger \,|\, 32^\dagger)$]
    [Colour Precision = $(8^\dagger \,|\, 16^\dagger \,|\, 24^\dagger \,|\, 32^\dagger)$]
    [Colour Selection Mode = (indexed† | direct†)]
    [Colour Value Extent = '*Direct Colour Value Pair*']
    [Index Precision = $(8^\dagger \,|\, 16^\dagger \,|\, 24^\dagger \,|\, 32^\dagger)$]
    [Integer Precision = $(8^\dagger \,|\, 16^\dagger \,|\, 24^\dagger \,|\, 32^\dagger)$]
    [Maximum Colour Index = '*non-negative integer*'†]
    [Real Precision = ({floating point format,9,23}†
                 | {floating point format,12,52}†
                  | {fixed point format,16,16}†
                  | {fixed point format,32,32}†)]
    [VDC Integer Precision = $(16^\dagger \,|\, 24^\dagger \,|\, 32^\dagger)$]
    [VDC Real Precision = ({floating point format,9,23}†
                  | {floating point format,12,52}†
                  | {fixed point format,16,16}†
                  | {fixed point format,32,32}†)]
    [VDC Type = (integer† | real†)] }

This attribute specifies the default values for the eleven CGM parameters listed in the attribute. If one or more of the parameters are missing, the following default values will be assumed: Colour Index Precision = 8; Colour Precision = 8; Colour Selection Mode = indexed; Colour Value Extent = ((0,0,0), (255,255,255)); Index Precision = 16; Integer Precision = 16; Maximum Colour Index = 63; Real Precision = {fixed point format,16,16}; VDC Integer Precision = 16; VDC Real Precision = {fixed point format,16,16}; VDC Type = integer.

The attribute transformation specification has the three parameters Clip Indicator, Clip Rectangle and VDC Extent. All parameters are optional:

transformation specification =
    {[Clip Indicator = (off[†] | on[†])]
    [Clip Rectangle = ' *VDC Pair* '[†]]
    [VDC Extent = ' *VDC Pair* '[†]]}

This attribute specifies the default values for the three CGM parameters listed in the attribute. If one or more of the parameters are missing, the following default values will be assumed: the default value for the parameter Transparency is on, the default value of the parameter is the value of the parameter VDC Extent, and the default value of the parameter VDC Extent is ((0,0), (1,1)).

The attribute transparency specification has the two parameters Auxiliary Colour and Transparency. Both parameters are optional:

transparency specification =
    {[Auxiliary Colour =
              (' *non-negative integer* '[†] | ' *Direct Colour Value* '[†])]
    [Transparency = (off[†] | on[†])]}

This attribute specifies the default values for the two CGM parameters listed in the attribute. If one or both of the parameters are missing, the following default values will be assumed: the default value for the parameter Transparency is on; the default value for the parameter Auxiliary Colour is 0 for the indexed colour mode and background for the direct colour mode.

The attribute colour representations has the parameters colour table specifications and Background Colour. The value of the parameter colour table specifications is a sequence whose elements consist of the two sub-parameters starting index and colour list. The value of the sub-parameter starting index is a non-negative integer. The value of the sub-parameter colour list is a set whose elements are *Direct Colour Values*. The values of the sub-parameters are encoded according to the binary encoding scheme of the CGM Standard. Both parameters are optional:

colour representations =
   { [colour table specifications =
      ⟦ [starting index = '*non-negative integer*'
         colour list = {['*Direct Colour Value*'†]$^+$}]$^+$⟧]
      [Background Colour = '*Direct Colour Value*'†]}

The parameter Background Colour specifies a default value for the corresponding CGM parameter. If the parameter is missing, its default value is background.

The parameter colour table specifications provides a mapping from direct colour values to an index number in the colour table. For the value *background* the colour table index 0 is used and for the value *foreground* the colour table index 1 unless otherwise specified. The default value for this parameter is an empty sequence.

The attribute edge rendition has the eight parameters edge aspect source flags, edge bundle specifications, Edge Bundle Index, Edge Colour, Edge Type, Edge Visibility, Edge Width and Edge Width Specification Mode.

The parameter edge aspect source flags has the sub-parameters edge colour asf, edge type asf and edge width asf. The values of these subparameters are either bundled or individual. The value of the parameter edge bundle specifications is a set whose elements consists of the two sub-parameters edge bundle representation and Edge Bundle Index. The sub-parameter edge bundle representation consists of the three subsubparameters Edge Colour, Edge Type and Edge Width. All parameters are optional. The attribute is therefore structured as shown on the next page.

This attribute specifies the default values for the presentation of the edges of the filled area primitives and, in particular, the default values for the CGM parameters Edge Bundle Index, Edge Colour, Edge Type, Edge Visibility, Edge Width and Edge Width Specification Mode. The parameter edge bundle specifications defines the initial edge representations at the beginning of the imaging process. A value should be assigned at least to the first five edge bundle indices; otherwise, the values are determined according to Fig. 68.

If one or more of the parameters are missing, the following default values will be assumed: edge aspect source flags = {edge colour asf = individual, edge type asf = individual, edge width asf =individual}; edge bundle specifications = {} (no specifications); Edge Bundle Index = 1; Edge Colour = 1 for indexed colour mode or Edge Colour = foreground for direct colour mode; Edge Type = 1; Edge Visibility = off; Edge Width = 1.0 for scaled mode or Edge Width = $\frac{1}{1000}$th of the length of the longest side

edge rendition =
    {[edge aspect source flags =
        {edge colour asf = (bundled | individual)
         edge type asf = (bundled | individual)
         edge width asf = (bundled | individual)}]
    [edge bundle specifications =
      {[edge bundle representation =
        {Edge Colour =
          ('*non-negative integer*'† | '*Direct Colour Value*'†)
         Edge Type = '*positive integer*'†
         Edge Width =
          ('*non-negative real*'† | '*non-negative VDC Value*'†)}
         Edge Bundle Index = '*positive integer*']$^+$}
      [Edge Bundle Index = '*positive integer*'†]]
    [Edge Colour =
      ('*non-negative integer*'† | '*Direct Colour Value*'†)]
    [Edge Type = '*positive integer*'†]
    [Edge Visibility = (off† | on†)]
    [Edge Width =
      ('*non-negative real*'† | '*non-negative VDC Value*'†)]
    [Edge Width Specification Mode = (absolute† | scaled†)]}

Edge Bundle Index	Edge Type	Edge Colour		Edge Width	
		indexed	direct	scaled	absolute
1	1 (solid)	1	foreground	1.0	$\frac{1}{1000}$th VDC size
2	2 (dash)	1	foreground	1.0	$\frac{1}{1000}$th VDC size
3	3 (dot)	1	foreground	1.0	$\frac{1}{1000}$th VDC size
4	4 (dash-dot)	1	foreground	1.0	$\frac{1}{1000}$th VDC size
5	5 (dash-dot-dot)	1	foreground	1.0	$\frac{1}{1000}$th VDC size
Note: The VDC size is the longest side of the default VDC extent					

Fig. 68: Default values for the edge bundle representations

of the VDC extent for absolute mode; Edge Width Specification Mode = scaled.

It should be noted that only those values are permitted for the parameter Edge Type which are either defined in the CGM Standard or which have been registered according to the procedure defined in this Standard; private values are not allowed.

The attribute filled area rendition has the ten parameters fill aspect source flags, fill bundle specifications, pattern table specifications, Fill Bundle Index, Fill Colour, Fill Reference Point, Hatch Index, Interior Style, Pattern Index and Pattern Size. The parameter fill aspect source flags has the four sub-parameters fill colour asf, hatch index asf, interior style asf and pattern index asf whose values are either bundled or individual. The value of the parameter fill bundle specifications is a set whose elements consist of the sub-parameters fill bundle representation and Fill Bundle Index. The value of the sub-parameter fill bundle representation consists of the four subsub-parameters Fill Colour, Hatch Index, Interior Style and Pattern Index.

The value of the parameter pattern table specifications is a set whose elements consist of the sub-parameters colour, local colour precision, nx, ny and pattern table specifications. The value of the subsub-parameter colour is either a sequence of non-negative integers or a sequence of *Direct Colour Values*. The number of elements in this sequence is equal to the product of the values of the subsub-parameters nx (number of columns of the pattern) and ny (number of rows of the pattern) since each cell of the pattern must have an associated colour value. The value of the sub-parameter local colour precision is either 0, 1, 2, 4, 8, 16, 24 or 32. The values of the subsub-parameters nx, ny and pattern table index are positive integers. All parameters are optional. The attribute is therefore structured as shown on the next page.

This attribute specifies the default values for the presentation of the interior of the filled area primitives and, in particular, the default values for the CGM parameters Fill Bundle Index, Fill Colour, Fill Reference Point, Hatch Index, Interior Style, Pattern Index and Pattern Size. The parameter fill bundle specifications defines the initial bundle representations at the beginning of the imaging process. The parameter pattern table specifications can be used to supply entries for the pattern table. A value should be assigned at least to the first five fill bundle indices; otherwise, the values are determined according to Fig. 69.

If one or more of the parameters are missing, the following default values will be assumed: fill aspect source flags = {fill colour asf = individual, hatch index asf = individual, interior style asf = individual, pattern index asf =individual}; fill bundle specifications = {} (no specifications); pattern table specifications = {} (no specifications); Fill Bundle Index = 1; Fill Colour = 1 for indexed mode or Fill Colour = foreground for direct mode; Fill Reference Point = first corner of the default VDC extent; Hatch In-

filled area rendition =
  {[fill aspect source flags =
    {fill colour asf = (bundled | individual)
    hatch index asf = (bundled | individual)
    interior style asf = (bundled | individual)
    pattern index asf = (bundled | individual)}]
  [fill bundle specifications =
    {[fill bundle representation =
      {Fill Colour =
        ('*non-negative integer*'† | '*Direct Colour Value*'†)
       Hatch Index = '*positive integer*'†
       Interior Style = (empty† | hatch† | hollow† | pattern† | solid†)
       Pattern Index = '*positive integer*'†}
     Fill Bundle Index = '*positive integer*']$^+$}]
  [pattern table specifications =
    {[colour =
    ([['*non-negative integer*'†]$^+$] | [['*Direct Colour Value*'†]$^+$])
    local colour precision = (0† | 1† | 2† | 4† | 8† | 16† | 24† | 32†)
    nx = '*positive integer*'†
    ny = '*positive integer*'†
    pattern table index = '*positive integer*'†]$^+$}]
  [Fill Bundle Index = '*positive integer*'†]
  [Fill Colour = ('*non-negative integer*'† | '*Direct Colour Value*'†)]
  [Fill Reference Point = '*VDC Pair*'†]
  [Hatch Index = '*positive integer*'†]
  [Interior Style = (empty† | hatch† | hollow† | pattern† | solid†)]
  [Pattern Index = '*positive integer*'†]
  [Pattern Size = {height vector x component = '*VDC Value*'†
               height vector y component = '*VDC Value*'†
               width vector x component = '*VDC Value*'†
               width vector y component = '*VDC Value*'†}]}

dex = 1; Interior Style = hollow; Pattern Index = 1; Pattern Size = {height vector x component = 0, height vector y component = height of the VDC extent, width vector x component = width of the VDC extent, width vector y component = 0}.

    It should be noted that only those values are permitted for the parameter Hatch Index which are either defined in the CGM Standard or which have been registered according to the procedure defined in this Standard; private values are not allowed.

Filled Area Bundle Index	Interior Style	Hatch Index	Pattern Index	Fill Colour indexed	direct
1	hollow	1 (horizontal parallel lines)	1	1	foreground
2	hatch	1 (horizontal parallel lines)	1	1	foreground
3	hatch	2 (vertical parallel lines)	1	1	foreground
4	hatch	3 (positiv slope parallel lines)	1	1	foreground
5	hatch	4 (negativ slope parallel lines)	1	1	foreground

Fig. 69: Default values for the filled area bundle representations

The attribute line rendition has the seven parameters line aspect source flags, line bundle specifications, Line Bundle Index, Line Colour, Line Type, Line Width and Line Width Specification Mode. The parameter line aspect source flags has the sub-parameters line colour asf, line type asf and line width asf. The values of these sub-parameters are either bundled or individual. The value of the parameter line bundle specifications is a set whose elements consists of the two sub-parameters line bundle representation and Line Bundle Index. The sub-parameter line bundle representation consists of the three subsub-parameters Line Colour, Line Type and Line Width. All parameters are optional. The attribute is therefore structured as shown on the next page.

This attribute specifies the default values for the presentation of the lines and, in particular, the default values for the CGM parameters Line Bundle Index, Line Colour, Line Type, Line Width and Line Width Specification Mode. The parameter line bundle specifications defines the initial line representations at the beginning of the imaging process. A value should be assigned at least to the first five line bundle indices; otherwise, the values are determined according to Fig. 70.

If one or more of the parameters are missing, the following default values will be assumed: line aspect source flags = {line colour asf = individual, line type asf = individual, line width asf =individual}; line bundle specifications = {} (no specifications); Line Bundle Index = 1; Line Colour = 1 for indexed colour mode or Line Colour = foreground for direct colour mode; Line Type = 1; Line Width = 1.0 for scaled mode or Line Width = $\frac{1}{1000}$th of the length of the longest side of the default VDC extent for absolute mode; Line Width Specification Mode = scaled.

line rendition =
  {[line aspect source flags =
     {line colour asf = (bundled | individual)
      line type asf = (bundled | individual)
      line width asf = (bundled | individual)}]
    [line bundle specifications =
     {[line bundle representation =
       {Line Colour =
         (*'non-negative integer'*[†] | *'Direct Colour Value'*[†])
        Line Type = *'positive integer'*[†]
        Line Width =
         (*'non-negative real'*[†] | *'non-negative VDC Value'*[†])}
       Line Bundle Index = *'positive integer'*][+]}]
    [Line Bundle Index = *'positive integer'*[†]]
    [Line Colour =
     (*'non-negative integer'*[†] | *'Direct Colour Value'*[†])]
    [Line Type = *'positive integer'*[†]]
    [Line Width =
     (*'non-negative real'*[†] | *'non-negative VDC Value'*[†])]
    [Line Width Specification Mode = (absolute[†] | scaled[†])]}

Line Bundle Index	Line Type	Line Colour		Line Width	
		indexed	direct	scaled	absoloute
1	1 (solid)	1	foreground	1.0	$\frac{1}{1000}$th VDC size
2	2 (dash)	1	foreground	1.0	$\frac{1}{1000}$th VDC size
3	3 (dot)	1	foreground	1.0	$\frac{1}{1000}$th VDC size
4	4 (dash-dot)	1	foreground	1.0	$\frac{1}{1000}$th VDC size
5	5 (dash-dot-dot)	1	foreground	1.0	$\frac{1}{1000}$th VDC size

Note: The VDC size is the longest side of the default VDC extent

Fig. 70: Default values for the line bundle representations

It should be noted that only those values are permitted for the parameter Line Type which are either defined in the CGM Standard or which have been registered according to the procedure defined in this Standard; private values are not allowed.

The attribute marker rendition has the seven parameters marker aspect source flags, marker bundle specifications, Marker Bundle Index, Marker Colour, Marker Size, and Marker Size Specification Mode and Marker Type.

The parameter marker aspect source flags has the sub-parameters marker colour asf, marker size asf and marker type asf. The values of these sub-parameter are either bundled or individual. The value of the parameter marker bundle specifications is a set whose elements consists of the two sub-parameters marker bundle representation and Marker Bundle Index. The sub-parameter marker bundle representation consists of the three subsub-parameters Marker Colour, Marker Size and Marker Type. All parameters are optional:

marker rendition =
  {[marker aspect source flags =
        {marker colour asf = (bundled | individual)
        marker size asf = (bundled | individual)
        marker type asf = (bundled | individual)}]
   [marker bundle specifications =
        {[marker bundle representation =
               {Marker Colour = ('$non\text{-}negative\ integer$'[†]
                    | '$Direct\ Colour\ Value$'[†])
             Marker Type = '$positive\ integer$'[†]
             Marker Size = ('$non\text{-}negative\ real$'[†]
                  | '$non\text{-}negative\ VDC\ Value$'[†])}
          Marker Bundle Index = '$positive\ integer$']$^+$}]
   [Marker Bundle Index = '$positive\ integer$'[†]]
   [Marker Colour = ('$non\text{-}negative\ integer$'[†]
          | '$Direct\ Colour\ Value$'[†])]
   [Marker Size = ('$non\text{-}negative\ real$'[†]
          | '$non\text{-}negative\ VDC\ Value$'[†])]
   [Marker Size Specification Mode = (absolute[†] | scaled[†])]
   [Marker Type = '$positive\ integer$'[†]]}

This attribute specifies the default values for the presentation of the marker primitives and, in particular, the default values for the CGM parameters Marker Bundle Index, Marker Colour, Marker Type, Marker Size and Marker Size Specification Mode. The parameter marker bundle specifications defines the initial marker representations at the beginning of the imaging process. A value should be assigned at least to the first five marker bundle indices; otherwise, the values are determined according to Fig. 71.

Marker Bundle Index	Marker Type	Marker Colour		Marker Size	
		indexed	direct	scaled	absoloute
1	1 (dot)	1	foreground	1.0	$\frac{1}{100}$th VDC size
2	2 (plus)	1	foreground	1.0	$\frac{1}{100}$th VDC size
3	3 (asterisk)	1	foreground	1.0	$\frac{1}{100}$th VDC size
4	4 (circle)	1	foreground	1.0	$\frac{1}{100}$th VDC size
5	5 (cross)	1	foreground	1.0	$\frac{1}{100}$th VDC size

Note: The VDC size is the longest side of the default VDC extent

Fig. 71: Default values for the marker bundle representations

If one or more of the parameters are missing, the following default values will be assumed: marker aspect source flags = {marker colour asf = individual, marker size asf = individual, marker type asf =individual}; marker bundle specifications = {} (no specifications); Marker Bundle Index = 1; Marker Colour = 1 for indexed colour mode or Marker Colour = foreground for direct colour mode; Marker Size = 1.0 for scaled mode or Marker Size = $\frac{1}{100}$th of the length of the longest size of the default VDC extent for absolute mode; Marker Size Specification Mode = scaled; Marker Type = 3 (asterisk).

It should be noted that only those values are permitted for the parameter Marker Type which are either defined in the CGM Standard or which have been registered according to the procedure defined in this Standard; private values are not allowed.

The attribute text rendition has the 17 parameters text aspect source flags, text bundle specifications, Alternate Character Set Index, Font List, Character Coding Announcer, Character Expansion Factor, Character Height, Character Orientation, Character Set Index, Character Set List, Character Spacing, Text Alignment, Text Font Index, Text Bundle Index, Text Colour, Text Path and Text Precision. The parameter text aspect source flags has the five sub-parameters character expansion factor asf, character spacing asf, text colour asf, text font asf and text precision asf whose values are either bundled or individual. The value of the parameter text bundle specifications is a set whose elements consist of the two sub-parameters text bundle representation and Text Bundle Index. The sub-parameter text bundle representation consists of the five subsub-parameters Character Expansion Factor, Character Spacing, Text Colour, Text Font Index and Text

Precision. All parameters are optional. The attribute is therefore structured as shown on the next page.

This attribute specifies the default values for the presentation of text and, in particular, the default values for the CGM parameters Alternate Character Set Index, Font List, Character Coding Announcer, Character Expansion Factor, Character Height, Character Orientation, Character Set Index, Character Set List, Character Spacing, Text Alignment, Text Font Index, Text Bundle Index, Text Colour, Text Path and Text Precision. A value should be assigned at least to the first two text bundle indices; otherwise, the values are determined according to Fig. 72.

Text Bundle Index	Font Index	Text Precision	Character Expansion Factor	Character Spacing	Text Colour indexed	direct
1	1	string	1.0	0.0	1	foreground
2	1	character	0.7	0.0	1	foreground

Fig. 72: Default values for the text bundle representations

If one or more of the parameters are missing, the following default values will be assumed: text aspect source flags = {character expansion factor asf = individual, character spacing asf = individual, text colour asf = individual, text font asf = individual, text precision asf = individual)}; text bundle specifications = {} (no specifications); Alternate Character Set Index = 1; Font List = list with one element denoting the registered name of a font which can represent the nationality-independent character subset of ISO 646; Character Coding Announcer = basic 7-bit; Character Expansion Factor = 1.0; Character Height = $\frac{1}{100}$th of the length of the longest side of the VDC extent; Character Orientation = $[\![(0,1), (1,0)]\!]$; Character Set Index = 1; Character Set List = {character set type = 94-character sets, designation sequence tail = designation sequence tail that is registered for a character set which includes the nationality-independent character subset of ISO 646 in the positions specified in ISO 646}; Character Spacing = 0.0; Text Alignment = {continuous horizontal alignment = normal horizontal, vertical alignment = normal vertical}; Text Font Index = 1; Text Bundle Index = 1; Text Colour = 1 for indexed colour mode or Text Colour = foreground for direct colour mode; Text Path = right; Text Precision = string.

It should be noted that only those values are permitted for the parameters Character Coding Announcer, designation sequence tail and Font

text rendition =
  {[text aspect source flags =
    {character expansion factor asf = (bundled | individual)
    character spacing asf = (bundled | individual)
    text colour asf = (bundled | individual)
    text font asf = (bundled | individual)
    text precision asf = (bundled | individual)}]
  [text bundle specifications =
    {[text bundle representation =
      {Character Expansion Factor = '*positive real*'†
      Character Spacing = '*real*'†
      Text Colour =
        ('*non-negative integer*'† | '*Direct Colour Value*'†)
      Text Font Index = '*positive integer*'†
      Text Precision = (character† | string† | stroke†)}
      Text Bundle Index = '*positive integer*']+}]
  [Alternate Character Set Index = '*positive integer*'†]
  [Font List = '*Registered Font Names List*'†]
  [Character Coding Announcer =
    (basic 7-bit† | basic 8-bit† | extended 7-bit† | extended 8-bit†)]
  [Character Expansion Factor = '*positive real*'†]
  [Character Height = '*non-negative VDC value*'†]
  [Character Orientation = [['*VDC Pair*'† '*VDC Pair*'†]]]
  [Character Set Index = '*positive integer*'†]
  [Character Set List =
    {character set type =
      (94-character sets† | 94-character multibyte sets†
      | 96-character sets† | 96-character multibyte sets†
      | complete code†)
    designation sequence tail =
      '*Registered Designation Sequence Tail*'†}]
  [Character Spacing = '*real*'†]
  [Text Alignment =
    {continuous horizontal alignment = '*real*'†
    continuous vertical alignment = '*real*'†
    horizontal alignment = (centre† | continuous horizontal†
                  | left† | normal horizontal† | right†)
    vertical alignment = (base† | bottom† | continuous vertical
                  | cap† | half† | normal vertical† | top†)}]
  [Text Font Index = '*positive integer*'†]
  [Text Bundle Index = '*positive integer*'†]
  [Text Colour =
    ('*non-negative integer*'† | '*Direct Colour Value*'†)]
  [Text Path = (down† | left† | right† | up†)]
  [Text Precision = (character† | string† | stroke†)]]}

List which are either standardized or registered; private values are not allowed.

8.2.3   Other Attributes for Geometric Graphics

Part 8 of the Standard defines also the values of three attributes introduced in Part 2 – content architecture class, content information and type of coding – if the attributes relate to geometric graphics content.

The value of the attribute content architecture class is an ASN.1 object identifier whose value is ⟦2 8 2 8 0⟧ for geometric graphics:

content architecture class = ⟦2 8 2 8 0⟧

This attribute can be specified for basic objects and basic object classes and indicates the content architecture class of content portions belonging to these objects or object classes. A value of ⟦2 8 2 8 0⟧ specifies that the geometric picture is in formatted processable form. Other content architecture classes are not possible for geometric graphics content.

The value of the attribute type of coding which can be specified for content portions, is an ASN.1 object identifier whose value is ⟦2 8 3 8 0⟧:

type of coding = ⟦2 8 3 8 0⟧

The attribute is specified for content portions containing a geometric picture. Since geometric graphics content is always encoded according to the binary encoding scheme of ISO 8632, Part 2, only this value is permitted. There are no further coding attributes for geometric graphics content.

The value of the attribute content information which can be specified for content portions, represents the actual geometric picture as an octet string:

content information = 'octet string'

The octet string is a *Computer Graphics Metafile* according to the ISO Standard 8632, encoded as specified in Part 3 of this Standard. The only restriction in relation to ISO 8632 is that the CGM may only contain one picture; according to ISO 8632, a CGM may contain more than one picture.

8.3   The Layout Process for Geometric Graphics

The layout process for geometric graphics is one of the three content layout processes described in the ODA Standard. It interacts with the document layout process to create the specific layout structure of a document. As for the other layout processes, the Standard does not specify in detail how the layout process for geometric graphics has to be carried out, but rather the result of this process is defined. This layout process is very similar to the scalable dimension method for raster graphics content as described in Sect. 7.3.2. The main difference is essentially that the aspect ratio is not defined by the ratio between the number of raster lines and pels per line but by the ratio between the vertical and horizontal size of the region of interest. However, in order to save the reader from a "translation" of the wording in Sect. 7.3.2 into the description applicable for geometric graphics, the layout process for geometric graphics content will be described in detail in this section.

The layout process for geometric graphics takes the content portion which is associated with a basic logical object or object class and contains a geometric picture in formatted processable form, and creates a basic layout object (block or basic page) where the picture is to be displayed. In particular, the dimensions (height and width) of the basic layout object are determined. They are forwarded to the document layout process which then then determines the precise positions of the basic layout object within the available area (see p. 170).

The document layout process informs the layout process for geometric graphics about the maximum dimensions available for the basic layout object. Within these constraints and based on the specifications made by the attributes that apply, the content layout process tries to determine the actual size of the basic layout object. If the picture fits into the available area, the content layout process can be successfully carried out.

If the picture does not fit, the content layout process fails and the document layout process may decide whether a larger area is available to start the content layout process again.

Before this process is described in detail, a few terms shall be introduced which are used later on.

h_{aa} shall indicate the horizontal size of the available area (see p. 170) which is given to the content layout process from the document layout process. The size is measured in scaled measurement units.

v_{aa} shall indicate the vertical size of the available area which is given to the content layout process from the document layout process. The size is measured in scaled measurement units.

h_{bl} shall indicate the horizontal size of the basic layout object (block) which is determined by the layout process for geometric graphics. The size is measured in scaled measurement units.

v_{bl} shall indicate the vertical size of the basic layout object which is determined by the layout process for geometric graphics. The size is measured in scaled measurement units.

ar shall indicate the aspect ratio of the region of interest. This value is the ratio between the width and the height of the region of interest if the attribute **picture orientation** has the value 0° or 180°, and the ratio between the height and the width if the attribute has the value 90° or 270° (see Fig. 65 and 66). For simplicity, we shall assume a picture orientation of 0° or 180° in the following.

Therefore, before the layout process for geometric graphics starts the values h_{aa}, v_{aa} and ar are always known and the values h_{bl} and v_{bl} are determined as a result of the layout process.

In the following, it is always assumed that the layout process is only concerned with that part of picture which is specified by the attribute **region of interest specification**, i.e., the clipped-off part of the picture is "invisible" for the layout process.

The determination of the values h_{bl} and v_{bl} depends on the value of the attribute **picture dimensions**. According to the four possible parameters of this attribute, the following four procedures can be distinguished:

1.   The parameter **width controlled** is specified for the attribute **picture dimensions** and the sub-parameters **minimum width** and **preferred width** indicate the minimum and maximum (preferred) width of the picture. The values h_{bl} and v_{bl} are determined in such a way that

   – h_{bl} is chosen as large as possible,
   –   the ratio between width and height of the picture is maintained and
   –   the basic layout object fits into the available area.

   In other words, the maximal h_{bl} is determined which satisfies the following three conditions:

   (1)   **minimum width** $\leq h_{bl} \leq \min(h_{aa},$ **preferred width**$)$

   (2) $v_{bl} \leq v_{aa}$

   (3) $\dfrac{h_{bl}}{v_{bl}} = ar$

(The function $\min(a, b)$ shall evaluate to the minimum of a and b.) It may, however, be possible that no values h_{bl} and v_{bl} can be found which fulfill these conditions.

For a better understanding consider the example in Fig. 73.

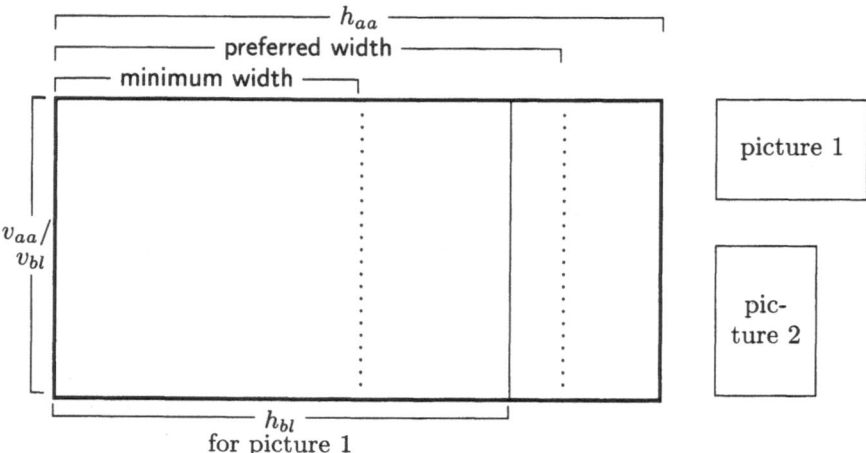

Fig. 73: Scalable dimension method for the parameter width controlled

The area with the bold border indicates the available area which is given from the document layout process to the content layout process for geometric graphics. The ratio between width and height for this area is $2 : 1$. The values of the parameters minimum width and preferred width of the attribute picture dimensions are shown graphically. The minimum width is the half, the preferred width $\frac{5}{6}$ of the width of the available area. In particular, the minimum width equals the value v_{aa} and the preferred width is $\frac{5}{3}v_{aa}$.

Beneath the available area two pictures are shown where the aspect ratio of the first is $3 : 2$ and of the second $2 : 3$. The situation at the start of the layout process can therefore be summarized as follows:

$$v_{aa} = x \text{ SMU} \quad \text{(the actual value of } x \text{ is not important for the example)}$$

$$h_{aa} = 2x \text{ SMU}$$

$$\text{minimum width} = x \text{ SMU}$$

$$\text{preferred width} = \frac{5}{3}x \text{ SMU}$$

$$ar = \frac{3}{2} \quad \text{for picture 1}$$

$$ar = \frac{2}{3} \quad \text{for picture 2}$$

For picture 1, the layout process will compute the value of v_{bl} equal to the height of the available area ($v_{bl} = v_{aa} = x$ SMU) and the value of h_{bl} equal to one and a half times the height ($h_{bl} = \frac{3}{2}v_{bl} = \frac{3}{2}x$ SMU). It can be easily checked that this is the largest value of h_{bl} which satisfies the conditions (1), (2) and (3).

For picture 2, the layout process can determine no values of h_{bl} and v_{bl} which satisfy all three conditions since the picture is a factor of $\frac{3}{2}$ higher than wide, i.e., the value of v_{aa} should be at least this factor larger than the minimum width in order to satisfy condition (2). Therefore, the layout process would fail for this picture.

For an aspect ratio greater than $\frac{5}{3}$, the layout process would use the preferred height (h_{bl} = preferred height), whereas the vertical size of the available area would not be needed completely ($v_{bl} < v_{aa}$).

2. The parameter height controlled is specified for the attribute picture dimensions and the sub-parameters minimum height and preferred height indicate the minimum and maximum (preferred) height of the picture.

The values h_{bl} and v_{bl} are determined in such a way that

- v_{bl} is chosen as large as possible,
- the ratio between width and height of the picture is maintained and
- the basic layout object fits into the available area.

In other words, the maximal v_{bl} is determined which satisfies the following three conditions:

(4) minimum height $\leq v_{bl} \leq \min(v_{aa}, $ preferred height$)$

(5) $h_{bl} \leq h_{aa}$

(6) $\dfrac{h_{bl}}{v_{bl}} = ar$

It may, however, be possible that no values h_{bl} and v_{bl} can be found which fulfill these conditions, i.e., the layout process may fail.

For a better understanding consider the example in Fig. 74.

The area with the bold border indicates the available area which is given from the document layout process to the content layout process for geometric graphics. The ratio between width and height for this area is 2 : 3. The values of the parameters minimum height and preferred height of the attribute picture dimensions are shown graphically. The minimum height is $\frac{2}{3}$ and the preferred height $\frac{8}{9}$ of the height of

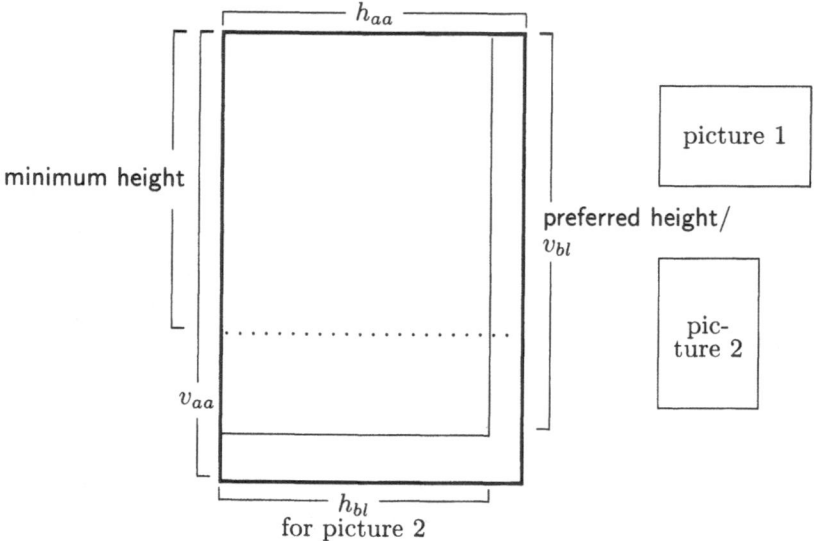

Fig. 74: Scalable dimension method for the parameter **height controlled**

the available area. In particular, the minimum height is equal to the value of h_{aa} and the preferred height is $\frac{4}{3}h_{aa}$.

Beneath the available area, two pictures are shown where the aspect ratio of the first is $3 : 2$ and of the second $2 : 3$.

The situation at the start of the layout process can therefore be summarized as follows:

$$h_{aa} = x \text{ SMU} \quad \text{(the actual value of } x \text{ is not important for the example)}$$
$$v_{aa} = \tfrac{3}{2}x \text{ SMU}$$
$$\text{minimum height} = x \text{ SMU}$$
$$\text{preferred height} = \tfrac{4}{3}x \text{ SMU}$$
$$ar = \tfrac{3}{2} \quad \text{for picture 1}$$
$$ar = \tfrac{2}{3} \quad \text{for picture 2}$$

For picture 2, the layout process will compute the value of v_{bl} equal to the preferred height ($v_{bl} = v_{aa} = \frac{4}{3}h_{aa}$) and the value of h_{bl} equal to $\frac{8}{9}h_{aa}$ since the following relation holds for the aspect ratio:

$$\frac{h_{bl}}{v_{bl}} = \frac{\frac{8}{9}h_{aa}}{\frac{4}{3}h_{aa}} = \frac{2}{3} = ar$$

For picture 1, the layout process can determine no values of h_{bl} and v_{bl} which satisfy all three conditions (4), (5) and (6) since the picture is a factor of $\frac{3}{2}$ higher than wide, i.e., the value of h_{aa} should be at least this factor greater than the minimum width in order to satisfy condition (5). Therefore, the layout process would fail for this picture.

3. The parameter **area controlled** is specified for the attribute **picture dimensions** and the sub-parameters **minimum height** and **preferred height** indicate the minimum and maximum (preferred) height of the picture and the sub-parameters **minimum width** and **preferred width** the minimum and maximum (preferred) width of the picture. Furthermore, the sub-parameter **aspect ratio flag** has either the value **fixed** or **variable**. Firstly, the following two conditions have to be satisfied:

   (7)   **minimum height** $\leq v_{aa}$
   (8)   **minimum width** $\leq h_{aa}$

This means that the available area must be large enough two contain the picture. The further processing method depends on the value of the sub-parameter **aspect ratio flag**.

If its value is **variable**, a distortion of the aspect ratio of the picture is allowed. In this case, as long as conditions (7) and (8) are satisfied, the layout process will always succeed with the values $h_{bl} =$ **preferred width** and $v_{bl} =$ **preferred height**. The picture will be distorted for

$$ar \neq \frac{\min(h_{aa}, \textsf{preferred width})}{\min(v_{aa}, \textsf{preferred height})}$$

If the sub-parameter **aspect ratio flag** has the value **fixed**, no distortion of the aspect ratio of the picture is permitted. In this case, the layout process determines the greatest possible values for h_{bl} and v_{bl} which satisfy the following conditions:

   (9)   **minimum width** $\leq h_{bl} \leq \min(h_{aa}, \textsf{preferred width})$
   (10)   **minimum height** $\leq v_{bl} \leq \min(v_{aa}, \textsf{preferred height})$
   (11) $\dfrac{h_{bl}}{v_{bl}} = ar$

An example for this method shall not be given since this method is essentially a combination of the two methods above and the examples given there should be sufficient to understand the layout process for the parameter **area controlled**.

4. The parameter **automatic** is specified for the attribute **picture dimensions**. In this case, the layout process tries to adjust the width of the picture to the width of the available area. The height of the picture will be such that the aspect ratio of the picture is observed, i.e., the following conditions apply to this method:

$$(12) \quad h_{bl} = h_{aa}$$

$$(13) \quad v_{bl} = \frac{h_{bl}}{ar}$$

This method can be considered a special case of the first method when taking there the values

minimum width = preferred width = h_{aa}

It should be noted that for all four methods the values of h_{bl} and v_{bl} must be integral multiples of one scaled measurement unit.

8.4   The Imaging Process for Geometric Graphics

The imaging process for geometric graphics content performs the actual "drawing" of the picture.

Firstly, an initialization process takes place which assigns to all CGM parameters the values specified by the presentation attributes, for example, the width of lines or the character fonts for text in a picture. In particular, this initialization process takes the attributes line rendition, marker rendition, text rendition, filled area rendition, edge rendition and colour representations into account. In other words, the evaluation of these presentation attributes has the same effect as if the corresponding CGM parameter had been explicitly specified at the beginning of the CGM.

Afterwards, the CGM, i.e., the value of the attribute content information, is evaluated and the picture is generated for a particular presentation device. The generation process is, of course, hardware-dependent and therefore not specified in the ODA Standard.

If the element SCALING MODE appears within the CGM, the imaging process ignores it since the aspect ratio and the required dimensions have already been determined by the layout process for geometric graphics content. Furthermóre, the picture descriptor element BACKGROUND

COLOUR which may appear within the CGM is ignored if for the basic layout object to which the picture belongs the attribute **colour** has the value **colourless** and the attribute **transparency** the value **transparent**.

9 Part 10: Formal Specifications

This Part of the ODA Standard defines a formal description technique based on a mathematical methodology and applies this technique to give a formal specification of the documents structures defined in Part 2 of the Standard.

Part 10 of ISO 8613 was published in 1991, i.e., about two years after the first publication of the other Parts of the Standard. Part 10 will be extended to contain also formal specifications of the document profile and the different content architectures. The publication of these Addenda is expected in late 1991 or 1992.

9.1 The Goal of the Formal Specifications

In the last few years, it has been widely accepted within ISO and other standard setting bodies that the continuously increasing complexity of standards, in particular in the field of information technology, requires advanced methods for the development and implementation of standards. Formal Description Techniques (FDTs), with precise and unambiguous syntax and semantics, usually based on mathematical means, are considered an important approach towards such advanced methods.

The application of a FDT can improve the quality of standards, since a formal method is more likely to reveal errors or loopholes in specifications than a natural language description.

These were in principle also the reasons which led to the development of the Formal Specifications of the ODA Standard (FODA). The aims of FODA are to provide

- a basis for implementations of the Standard,
- a tool for the verification of conforming system and

– a reference point for examining future extensions and revisions to the Standard.

Part 10 does not add any new technical specifications to the ODA Standard: it only "repeats" technical specifications from other Parts of the Standard in a rigorous and unambiguous formal notation. Nevertheless, a reading of Part 10 may be worthwhile since it may give some deeper insights into certain aspects of the ODA Standard.

The development of FODA was started in 1985 when the other Parts of the Standard which use natural English for their technical specifications were already well advanced. At the beginning, the work concentrated on Part 2 of ISO 8613 since this part plays a central role: the *document structures* can be regarded as the kernel of the whole Standard, supporting the other parts, for instance, the different content types of an ODA document. Part 10 as initially published contains only the formal specifications of these documents structures but the formal specifications of the document profile and the different content architectures is already well advanced and is expected for publication soon.

Ideally, a formal specification of the technical contents of a Standard and their natural language description should be progressed in parallel since the formal specifications would rather likely improve the natural language text. Nevertheless, the *a posteriori* approach which was undertaken for the formal specifications of the ODA Standard proved quite useful. The development of FODA revealed a significant number of errors, ambiguities and loopholes in the natural language text. A part of these problems could be resolved before the publication of Parts 1 to 8 of ISO 8613, others were (and still are) corrected afterwards by means of the ISO defect handling procedures.

Though the primary goal of the FODA development was a "checking" of the natural language specifications by a translation into a formal language, it turned out in the meantime that FODA has also several useful applications in practice. In particular, the development of conformance testing software which tests whether a given ODA stream conforms to the specifications of ISO 8613, can profit rather straightforward from FODA. This is explained in some detail in Sect. 9.4.

9.2   The Formal Description Technique

The FDT used for FODA is called the *Information Modelling by Composition Language* (IMCL); it is defined in clause 4 of Part 10. This language has its mathematical basis in elementary set theory and first-order predicate logic, augmented with some features that allow the modelling of those structures which are encountered in many areas of computer science and, in particular, in ODA documents. It should be rather easy to learn for the expected readership, namely implementors of ODA systems or ODA-related software, who will generally have a background in computer science and therefore understand set theory and predicate logic.

This section gives a short introduction to IMCL, though a complete survey is beyond the limits of this book.

IMCL is based on the the following concepts:

1. The "universe" of IMCL is a non-empty set of entities of the following kinds:

    - *constructs*,
    - *spots*,
    - *spotsets* (i.e., sets of spots) and
    - the entity **UNDEF** ("undefined").

2. A certain set of functions from the universe to the universe is defined, i.e., operators on entities of the universe.
3. A certain set of predicates in the universe is defined, i.e., predicates on entities of the universe.
4. A *construct* is an information object which is either

    - an *atomic construct* (*atom*) or
    - a *composite construct*.

5. A *composite construct* is either

    - a *collection* which is an unordered set of component constructs,
    - a *catenation* which is a sequence of component constructs, or
    - a *nomination* which is a function that can be regarded as an unordered set of ordered pairs where each pair consists of a *name* and a *value*.

The special terminology for composite constructs is to distinguish them from other sets, sequences or functions in mathematical terminology.

Atomic constructs

Atomic constructs are either numbers or special atoms with application
defined semantics. They are syntactically denoted by character strings.
For instance, 314 is a number, 'a' or 'ODA' are special atoms. Character
strings denoting special atoms are enclosed by single quotes. It should
be noted that 'ODA' is considered atomic; it is *not* a sequence of three
characters and cannot be decomposed.

In the formal specifications of ISO 8613, the following atomic constructs appear:

- numbers (integers and real numbers), for example as values for certain
  attributes;
- attribute names, parameters names, sub-parameter names, etc., for
  example, 'protection' or 'object identifier';
- certain non-composite attribute values, parameter values, sub-parameter values, etc., for example 'protected' or 'null';
- certain entities which are referenced but not defined in ISO 8613, for
  example a date value which conforms to ISO 8601.

Composite constructs

For explicitly specifying the composition of composite constructs (collections, catenations and nominations) IMCL uses the following notation:

- Collections are denoted by enclosing their components in brack-
  ets and separating the elements by a semicolon. For instance,
  [314; 'a'; 'ODA'] is a collection with the three (atomic) components 314, 'a' and 'ODA'. An empty collection is denoted by [].
- Catenations are denoted by enclosing their components in brack-
  ets and surrounding their elements with an arrow. For example, [→'O'→'D'→'A'→] is a catenation consisting of the three
  (atomic) constructs 'O', 'D' and 'A'. An empty catenation is denoted by [→].
- Nominations are denoted by enclosing the ordered pairs in brack-
  ets and separating the pairs by a semicolon. The *name*-part and
  the *value*-part of each pair are separated by a colon. For instance,
  ['id': 7; 'std': 'ODA'] is a nomination consisting of two (*name*,
  value)-pairs; the first one has the name 'id' with the value 7 and
  the second one the name 'std' with the value 'ODA'. An empty
  nomination is denoted by [:].

Composite constructs can be arbitrarily nested using atoms, collections, catenations and nominations.

These IMCL constructs are sufficient to model any ODA document and, consequently, all structures in ODA documents. For example, the specific logical structure can be formally described as a collection whose elements are the constituents of the specific logical structure (see Sect. 3.1.1). A constituent such as a basic logical object or a presentation style can be formally described as a nomination whose ordered pairs are the attributes and their respective values. Any attribute value can be formally described using atoms, a collections, catenations and nominations.

The formal notation of a particular constituent of type *composite logical object* (see p. 45), having the attributes object identifier, object type, protection and subordinates, might be, for instance:

```
['object identifier': [→3→5→8→];
 'object type': 'composite logical object';
 'protection': 'protected';
 'subordinates': [→0→3→1→4→]]]
```

The composite logical object is modelled in IMCL as a nomination with the *name*-parts 'object identifier', 'object type', 'protection' and 'subordinates'. The *value*-part belonging to 'object identifier' is a catenation of three integers (see p. 72 and 88), the *value*-part belonging to 'object type' is the atom 'composite logical object' (see p. 88), the *value*-part belonging to 'protection' is the atom 'protected' (see p. 137), and the *value*-part belonging to 'subordinates' is a catenation of four integers (see p. 91).

In the above manner, any ODA document could be completely encoded using IMCL, i.e., IMCL can be considered an alternative data format to ODIF for the encoding of ODA documents. The IMCL encoding of an ODA document is both human-readable and machine-processable. Translations from the ODIF encoding to an IMCL encoding and vice versa are rather straightforward.

However, the purpose of the formal specifications of ISO 8613 is not primarily for the formal description of a specific ODA document but rather for the formal description of the general rules which hold for ODA documents, for instance, which value ranges are permitted for attributes, which attributes are mandatory, non-mandatory or defaultable for constituents and which sets of constituents are permitted or required in documents of a given document architecture class.

Spots and spotsets

In order to be able to address components in composite structures, IMCL uses the concept of a *spot*. This concept is an abstract counterpart for the intuitive idea associated with pointing into a structure at some position and saying "here". However, in general the "here" is not identified uniquely by the component construct as such (for instance in a word, the same letter may occur several times), but rather by the context in which it appears. To deal with the concept of "here" requires a way to identify contexts.

The concept of a spot allows the distinction to be made between a particular construct and its position within a composite construct of which it is a component. The character string $[\rightarrow\text{'d'}\rightarrow\text{'a'}\rightarrow\text{'t'}\rightarrow\text{'a'}\rightarrow]$ (a catenation), for example, has the component constructs 'd', 'a' and 't'. Whereas 'd' and 't' appear at one spot only, the 'a' appears at two spots, namely at the second and at the fourth position counted from the front end. Therefore, $[\rightarrow\text{'d'}\rightarrow\text{'a'}\rightarrow\text{'t'}\rightarrow\text{'a'}\rightarrow]$ has four component spots, but only three component constructs. If a construct is considered outside any context, it is said to be at its *ownspot*.

Spots are usually identified by selection criteria which are built on operators (see Section "Operators" below). However, a selection criterion need not be unique. Thus, it is more natural to deal with sets of spots (*spotsets*) rather than spots. Consequently, IMCL only contains expressions for *singleton spotsets* (spotsets with only one component) and not for individual spots.

Predicates

"Facts" in an IMCL application are usually expressed by predicates in first-order predicate logic. From predicate logic, IMCL uses the logical connectors *and*, *or*, *xor*, *iff* (if and only if) and *impl* (implies) and the logical quantifiers $\forall$ (for all) and $\exists$ (exists).

Several predicates, yielding either true or false, are already predefined, for instance:

IsAtom(t)   is true, if and only if (iff) t is an atomic construct;
IsInt(t)   is true, iff t is an integer;
IsCol(t)   is true, iff t is a collection;
IsCat(t)   is true, iff t is a catenation;
IsNom(t)   is true, iff t is a nomination;
$t_1 = t_2$ is true, iff the entities t_1 and t_2 are identical;

$t_1 \in t_2$ is true, iff t_1 is an element of the collection t_2.

Based on these basic predicates and operators additional predicates can be defined. For instance, a predicate IsNeNom ("is non-empty nomination") which applied to a construct *con* yields true if and only if *con* is a nomination but not the empty nomination, could be defined as:

$\forall$ *con* (IsNeNom(*con*) *iff* IsNom(*con*) *and* *con* $\neq$ [ ])

This could be read as: "An entity *con* is a non-empty nomination if and only if it is a nomination but not the empty nomination."

As another example, the concept that a basic logical object (like any other constituent in an ODA document) is a non-empty set of attributes modelled as a nomination could be expressed by the following predicate:

$\forall cst$ (IsBasicLogicalObject(*cst*) *iff* IsNeNom(*cst*))

This could be read as: "An entity *cst* is a basic logical object if and only if it is a non-empty nomination." (Of course, additional constraints must be satisfied for an entity *cst* to be a valid basic logical object).

Operators

Several operators (also called functions) are part of IMCL. For numbers the usual arithmetical operators $(+ - \times /)$ are defined. For collections the set operators $\cup$ (union), $\cap$ (intersection) and \ (set difference) identical to those in mathematical set theory are included in IMCL. For catenations the concatenation operator $//$ is defined.

Several further operators dealing with names, components and spotsets, are also part of the specification language, for instance:

$\hat{}t$ If t denotes a construct, then $\hat{}t$ ("ownspot of t") denotes the singleton spotset containing the ownspot of t.

         This operator transforms a construct into a singleton spotset. Usually, it is applied before using further operators, such as N t or C t, which are defined only for spotsets.

$t \bullet$ If t denotes a spotset containing no atom spots (spots with atoms), then $t \bullet$ ("next inwards") denotes the set of all spots which are immediately inward of the spots of the spotset t.

         This operator expands a spotset of composite components into a spotset of the immediate component spots of the composite component spots. This is most commonly applied to a singleton spotset.

C t If t denotes a singleton spotset, then C t denotes the compo-
                nent construct at the spot given by t.

N t If t denotes a singleton spotset of a spot that is a component
                of a nomination, then N t denotes the name construct of the
                component as it is within the nomination.

                N t is often used when constraining the values of at-
                tributes.

t.'name'   If t denotes a spotset of a nomination spot then t.'name'
                returns a spotset which is the set of all immediately inward
                spots (of the nomination spots) for which the name con-
                structs are 'name'.

                Usually, this is used when t is a singleton spotset of a
                nomination spot, thus, assuming that 'name' is a valid name
                construct within the nomination, then t.'name' returns a
                singleton spotset of the value stored under 'name'.

$x \hat{\in} t$ If t denotes a spotset then $x \hat{\in} t$ is true iff x is a singleton
                spotset of one of the spots in t.

                This predicate is most often used to specify properties of
                a spotset.

Using these operators additional operators can be defined. For example, an operator NAMS (read "name set") which returns all the *name*-
parts of the ordered pairs of a nomination, could be defined as follows:

$\forall n$ (NAMS(n) =
      IF IsNom(n) THEN [m | $\exists \, p \, \hat{\in} \, \hat{} n$. ($m$ = N p)] ELSE UNDEF)

This could be read as: "If an entity n is a nomination then the operator
NAMS returns the collection of those entities m, which are the *name*-
parts at the component spots p of the nomination n. Otherwise the value
UNDEF is returned."

9.3   The Formal Specifications of the Document Structures

The formal specifications of the document structures are given in clause 7
of ISO 8613, Part 10. They are essentially a single formula in first-order
predicate logic. The formula consists of sub-formulae which are connected
by *and*:

formula$_1$ _and_ formula$_2$ _and_ formula$_3$ _and_ ... formula$_n$

A formula is also called a _definition_ since it either defines a _concept_ used in the natural English description of the ODA Standard, or a so-called _subsidiary predicate_ or _subsidiary function_ which has been introduced for the sake of readability. The definitions are classified into four groups:

- The first group contains all formulae defining the _sets of constituents_ such as a whole document, a document body, a generic part or a generic logical description (see Fig. 14).
- The second group contains the formulae defining the _constituents_ such as a logical object class description, a logical object description, a content portion description, a layout style or a presentation style (see again Fig. 14). Also the constituents concerning the objects of the ODA layout structure such as page, frame and block are defined within this group.
- The third group contains all the formulae defining the structure and the value ranges of the _attributes_ which occur in the constituents.
- The last group contains _subsidiary definitions_ which are used in various definitions in the preceding three groups. For example, some constituents of an ODA document are embedded in hierarchical tree structures with complicated cross relationships which are rigorously defined within this group.

Each definition is introduced by a so-called semi-formal description. These descriptions give an explanation of the definitions to which they belong in English, making an extensive use of the terms of the specification method. The semi-formal descriptions do not belong to the formal specification itself, however, they are intended to help understand the formal definitions.

Though a complete survey of FODA cannot be given in this book a few examples may illustrate the principles.

The first example, belonging to the above mentioned first group of definitions, is the predicate "IsDocumentDescription" which is defined as shown in Fig. 75.

As can be seen, each definition has a unique number (2.3 for this predicate). For the sake of readability, the definitions are visually structured into a number of lines with the line numbers shown at the left-hand side. The line numbers are not part of the formal specifications but they are usually referenced in the semi-formal descriptions.

Furthermore, subscripts are often attached to opening and closing parentheses so the reader can quickly see the balancing structure of more

$$\boxed{\text{Definition 2.3}}$$

1 $\forall doc$
2 $(_0\,\text{IsDocumentDescription}(doc)\;\underline{iff}$
3    $\exists prof$
4    $(_1\,\text{IsDocumentProfilePart2}^{2.20}(prof)\;\underline{and}$
5     $(_2\,doc = [prof]\;\underline{or}\;\text{IsProcessable}^{2.4}(doc)\;\underline{or}$
6      $\text{IsFormattedProcessable}^{2.5}(doc)\;\underline{or}\;\text{IsFormatted}^{2.6}(doc)\;_2)\;\underline{and}$
7     $doc \neq [prof]\;\underline{impl}$
8     $(_3\,(_4\,\text{C}\;\widehat{}prof \cdot {}'\text{document architecture class}' = {}'\text{processable}'\;\underline{iff}$
9      $\text{IsProcessable}^{2.4}(doc)_4)\;\underline{and}$
10    $(_5\,\text{C}\;\widehat{}prof \cdot {}'\text{document architecture class}' = {}'\text{formatted processable}'\;\underline{iff}$
11     $\text{IsFormattedProcessable}^{2.5}(doc)_5)\;\underline{and}$
12    $(_6\,\text{C}\;\widehat{}prof \cdot {}'\text{document architecture class}' = {}'\text{formatted}'\;\underline{iff}$
13     $\text{IsFormatted}^{2.6}(doc)_6)\;\underline{and}$
14    $\forall cst \in doc$
15    $(_7\,{}'\text{resource}' \in \text{NAMS}^{1.18}(cst)\;\underline{impl}$
16     ${}'\text{resource document}' \in \overline{\text{NAMS}^{1.18}}(prof)\;_7)_3)_1)_0)$

Fig. 75: Definition of the predicate IsDocumentDescription

complicated expressions. Superscripts are attached to referenced predicates or functions of they are not predefined in IMCL but given somewhere else. For instance, in the example above the predicate "IsDocumentProfilePart2" is defined as definition 2.20. Again, the subscripts and superscripts are not an intrinsic part of the formal specifications but are only used for the sake of readability.

The semi-formal description belonging to this definition reads as follows:

"An entity *doc* is a document description if it is a document profile *prof* (line 5) or if it is an entity which is processable, formatted processable or formatted (5–6), according to the value of the document profile attribute 'document architecture class' (7–13), with a 'resource document' specified in the profile if any 'resource' is specified in the document (14–16)."

Of course, all "facts" expressed by this definition are already specified in Part 2 of the Standard (see, for instance, Sects. 3.3.1 and 3.3.2 and pp. 98, 186 and 187).

The predicate "IsBasicLogicalObjectDescription" is an example from the second group of definitions dealing with the formal specification of the constituents. This predicate is defined as shown in Fig. 76.

$$\boxed{\text{Definition 2.41}}$$

1 $\forall cst$

2 $(_0$ IsBasicLogicalObjectDescription(cst) *iff*

3   IsAttributeSet$^{2.62}(cst)$ *and*

4   NAMS$^{1.18}(cst) \supseteq$ ['application comments';   'bindings';

5   'content architecture class'; 'object identifier';

6   'object type';   'protection';

7   'user-readable comments';   'user-visible name'] *and*

8   NAMS$^{1.18}(cst) \subseteq$ ['application comments';   'bindings';

9   'content architecture class'; 'content generator';

10   'content portions';   'layout style';

11   'object class';   'object type';

12   'object identifier';   'presentation style';

13   'protection';   'user-readable comments';

14   'user-visible name'] *and*

15    $\forall a \,\hat{\in}\, {}^\wedge cst$.

16    $(_1 (_2$ N $a =$ 'object class' *impl* IsLogicalObjectClassId$^{2.81}$(C a $)_2$) *and*

17       $(_3$ N $a =$ 'object identifier' *impl* IsLogicalObjectId$^{2.78}$(C a $)_3$) *and*

18       $(_4$ N $a =$ 'object type' *impl*

19          $(_5$ C $a =$ 'basic logical object' *or* IsPlaceholder$^{1.19}$(C a $)_5)_4)_1$) *and*

20    $(_6$ NAMS$^{1.18}(cst) \cap$

21   ['content portions'; 'object class'; 'content generator'] $\neq$ [ ]$_6$) *and*

22    $(_7$ IsPlaceholder$^{1.19}$(C $^\wedge cst$. 'object type') *impl*

23   *not* IsPlaceholder$^{1.19}$(C $^\wedge cst$. 'object class')$_7)_0$)

Fig. 76: Definition of the predicate IsBasicLogicalObjectDescription

The semi-formal description belonging to this definition reads as follows:

"A basic logical object description is an attribute set which contains the attributes 'application comments', 'bindings', 'content architecture class', 'object identifier', 'object type', 'protection', 'user-readable comments', 'user-visible name' and, optionally, the attributes 'layout style', 'content portions', 'content generator', 'presentation style' and 'object class' (4–14). The attributes 'object class' and 'object identifier' have a logical object class identifier and a logical object identifier as respective values (16, 17). The attribute 'object type' has 'basic logical object' as its value (18, 19). At least one of the attributes 'object class', 'content portions' and 'content generator' must be specified for this constituent (20, 21). If the value of the attribute 'object type' is defaulted the value of the attribute 'object class' must be specified explicitly (22, 23)."

The predicate "IsAttributeSet" appearing in line 3 guarantees that the values of all attributes lie in their permitted value ranges. This predicate specifies also whether an attribute is defaultable or not by making use of the predicate "IsPlaceholder". The predicate "IsPlaceholder(v)" – it appears also in line 22 of the example – is true if and only if the attribute value v is defaulted and not specified explicitly on a given constituent. In particular, lines 22 and 23 of the definition above specify formally the Note 1 on p. 45 which applies to a basic logical object.

The predicate "IsMediumTypeValue" is an example from the third group of definitions dealing with the formal specification of attribute values. This predicate is defined as shown in Fig. 77.

| Definition 2.105 |

```
1   ∀v
2   (₀ IsMediumTypeValue(v) iff
3      IsNeNom^{1.2}(v) and
4      NAMS^{1.18}(v) = ['nominal page size'; 'side of sheet'] and
5      ∀a ê ˆv •
6      (₁ (₂ N a = 'nominal page size' impl
7         (₃ IsPlaceholder^{1.19}(C a) or IsPairOfPosInt^{1.9}(C a)₃)₂) and
8         (₄ N a = 'side of sheet' impl
9         (₅ IsPlaceholder^{1.19}(C a) or
10           C a ∈ ['recto'; 'verso'; 'unspecified']₅)₄)₁)₀)
```

Fig. 77: Definition of the predicate IsMediumTypeValue

The semi-formal description belonging to this definition reads as follows:

"The value of the attribute 'medium type' is a nomination of two components (3, 4). For the name 'nominal page size' the component is a pair of positive integers, the first one specifying the horizontal, the second one the vertical dimensions of the page (6, 7). For the name 'side of sheet' the component is either 'recto', 'verso' or 'unspecified' (8–10). The parameters are independently defaultable (7, 10)."

9.4   Applications of the Formal Specifications

Though the development of the formal specifications of the ODA Standard were started without a direct practical application in mind – except a "checking" of the natural language specifications – it turned out in the meantime that FODA can be applied with great benefit for the development of conformance testing software.

The reason for this is as follows: Conformance testing to a standard such as ISO 8613 is usually done by dedicated software and, as a consequence, developing conformance testing software needs an interpretation of the standard in question and a translation of its specifications into a programming language. Programming languages are – in contrast to natural English – formal languages with precise syntax and semantics.

It seems obvious that a translation of one formal language into another should be straightforward and much faster than the translation of an informal specification into a formal one, especially when dealing with such a complex standard as ISO 8613. Instead of an evaluation of the natural language specifications of the ODA Standard and the derivation of software specifications for conformance testing programs from this evaluation, it turned out to be much faster to take FODA as a basis for the specification of ODA conformance testing software.

The effort of translating FODA into conformance testing software depends largely on the programming language used. When using Prolog as the target programming language, a Canadian study showed that the translation is very easy and the resulting Prolog code resembles very much the original FODA specifications. This is due to the fact that IMCL and Prolog are both based on first-order predicate logic. When using procedural programming languages such as Pascal or C, the resulting code may look rather different from the original FODA specifications but a translation is nevertheless possible.

As mentioned already in Sect. 2.3, there are two fundamentally different aspects of conformance in the context of ODA:

– The first aspect addresses whether or not a given ODA *document*, encoded as an ODIF data stream, conforms to ISO 8613. In this case it has to be decided whether all the data elements within the data stream are such as defined in the Standard.

– The second aspect addresses whether or not a given ODA *implementation* conforms to ISO 8613. In this case it has to be decided whether an implementation can process (receive, transmit, edit, format, image, etc.) ODA documents as intended by the Standard. This means, for example, that the semantics of the attributes have been implemented correctly and, in particular, that the layout process has been implemented in such a way that it meets the specifications.

Concerning the conformance testing of ODA *documents*, encoded as ODIF data streams, it has to be checked if the data stream is a valid ASN.1 encoding as specified in ISO 8613, Part 5. Such a test can be implemented rather straightforwardly but such a syntactical test is only of some limited value since it cannot take into account the complex cross relationships between the constituents and attributes within an ODA document, according to the semantics of the Standard. These cross relationships, however, are covered by the FODA specifications i.e., evaluating all the formulae of FODA for a given test document will either result in "true" or "false" which decides whether or not the document conforms to ISO 8613.

It should be noted that such an approach for conformance testing has an additional advantage: The issues of the conformance test itself are rigorously and unambiguously defined, i.e., it is formally specified which properties of a document are tested and which are not, for example, because they are considered implementation-dependent. If conformance tests are based on informal specifications there exists the danger that unrevealed assumptions for specific features are implemented in the test software.

Testing the conformance of ODA *implementations* requires the specification of an appropriate test suite, i.e., a set of test documents. An Implementation Under Test (IUT) must be able to generate a specific test document (i.e., its ODIF data stream) or to process a given document in various predefined ways. A thorough test requires a large number of test documents, especially if additional constraints imposed by different Document Application Profiles have to be taken into account.

It is obvious that the set of test documents must be defined carefully so that a processing of these tests documents will provide a sufficient degree of confidence that the IUT is a conforming implementation.

FODA can also be used for the specification of such test documents, since it is a well structured and a highly systematic representation of the specifications of ISO 8613. Cross relations, for instance, between different attributes, are usually clearly visible in FODA and can often be identified more easily than by an inspection of the natural English version of the Standard. For example, if test cases for the correct implementation of the semantics of a given attribute are to be specified, FODA can be used in a straightforward manner for a systematic generation of suitable test documents.

In fact, several developers of conformance testing software for ODA use FODA as a basis for their developments. A detailed survey of these projects, however, is outside the scope of this book.

10 Annex: Modifications and Extensions of ISO 8613

Since its first publication in 1989 several modifications and extensions of ISO 8613 have taken place and additional changes are expected in the future. It should be noted that all changes to ISO 8613 are also implemented in the CCITT T.410 series of Recommendations, i.e., the ISO Standard and the CCITT Recommendations remain aligned.

Unfortunately, finding out the exact status of an ISO Standard at a particular point in time is not very easy. The publication of changes to a Standard by the ISO Central Secretariat in Geneva may take some time after such changes have been officially agreed within the ISO committees. Furthermore, purchasers of ISO Standards are usually not informed automatically about changes. Readers interested in the definitive version of the ODA Standard at a particular point in time should therefore contact their National Standards Bodies (AFNOR in France, ANSI in the USA, BSI in the UK, DIN in Germany, etc.) for the current status of the Standard.

This book reflects the status of the ODA Standard in Spring 1991, i.e., modifications and extensions that had been agreed by then are included in the text. Their official publication is expected in Summer 1991. This annex gives a short survey of these changes, in particular, for readers who have the original published version of the Standard, to understand some discrepancies between this book and their version of the Standard.

10.1 Technical Corrigenda

A number of changes – called *technical corrigenda* – have been made to correct errors or resolve ambiguities of the Standard. Only three of these changes shall be listed here, those in particular which lead to an obvious discrepancy between this book and the original published version of the Standard.

1. The attribute content type has been removed.

    The first version of Part 2 of the Standard contained the attribute content type which provided an alternative method to specify the content architecture class of content portions belonging to a basic object. This – in principle redundant – attribute had been introduced for compatibility with existing CCITT Recommendations. However, this could lead to problems with the defaulting mechanism if both content type and content architecture class had to be defaulted and the attribute content type was therefore removed. This required changes at several places in Parts 1, 2, 4 and 5 of the Standard.

2. The parameters logical object and layout object have been added to the attribute same layout object.

    Though the first version of Part 2 of the Standard used the term "parameter" in the description of the attribute same layout object, it did not, as required, specify the names of these parameters. This has been corrected and the description of the attribute has been modified accordingly.

3. The structure of the attribute character fonts has been changed.

    In the first version of Part 6 of the Standard the attribute character fonts was defined in an ambiguous way which, furthermore, did not match its ODIF representation. This was corrected by introducing the ten parameters primary font, first alternative font ... ninth alternative font with the two sub-parameters font size and font identifier into the attribute.

Further changes of this kind will be made if additional problems' arise.

10.2   Addendum on Document Application Profile Proforma and Notation

The originally published version of Part 1 of the Standard did not contain a precise notation for how Document Application Profiles (DAPs) should be written. It contained only a rather informal description of the contents of DAPs.

When the development of DAPs within ODA user groups commenced, the missing formal notation to be used for DAPs was considered a major problem. Therefore, the Addendum on Document Application Profile Proforma and Notation was developed. The contents of this Addendum to Part 1 of the Standard are described briefly in Sect. 2.2.3.

10.3   Addendum on Styles

As experiences with implementations of the ODA Standard showed, it is often the case that different layout styles and presentation styles are very similar. For instance, the difference between two styles X and Y may be that style Y specifies only one additional attribute or modifies only the value of one particular attribute compared to style X.

It was therefore considered a useful feature if a style could refer to another style to "inherit" the attributes specified there and thus reduce the amount of information that has to be stored in an ODA document. This led to the development of the Addendum on Styles which caused a few modifications in Parts 2 and 5 of the Standard.

In particular, the concept of a *root style* and a *derived style* was introduced and the attribute derived from was added. The technical specifications of this Addendum are reflected in Sects. 3.1.5 and 3.2.7.

10.4   Addendum on Alternate Representations

When developments of ODA related systems started, in particular, developments of converters between existing document processing systems and ODIF, the "fallback mechanism" provided by the attribute alternative representation, for implementations which could not handle all features of the ODA Standard, was considered too simple. For instance, a particular system may be able to handle character content and raster graphics content but not geometric graphics content. If a geometric picture appeared in a received data stream, the receiver could only be informed on the presence of the picture by means of the attribute alternative representation, but the picture itself could not be imaged.

On the other hand, a transformation of a geometric picture into a raster image on the originator's side could be rather straightforward and providing the recipient with a rasterized version of a geometric picture would be an advantage. Therefore, the Addendum on Alternate Representations was developed which added the concept of primary and alternative representations to the Standard and, in particular, the attributes alternative, primary and alternative feature sets of Parts 2 and 4. Some additional modifications were necessary in Parts 1 and 5.

The technical specifications of this Addendum are reflected in Sects. 3.2.5, 3.2.14 and 4.2.2.

10.5   Addendum on Security

The ODA Standard was developed for the interchange of documents in an
open systems environment, i.e., a document transmitted by an originator
will usually travel through an electronic network before it finally reaches
the intended recipient. The originator and recipient will usually not know
the intermediate places – such as mail servers – which the document
passes. This immediately raises questions relating to security aspects
and therefore the Addendum on Security was developed.

This Addendum added the concepts of the so-called *protected part* to
the Standard and, in particular, introduced the attributes enciphered, enciphered information, protected part identifier, sealed and sealed document
profile information to Part 2 and the attributes enciphered document profiles, enciphered profiles, ODA security label, post-enciphered body parts,
post-enciphered document body parts, post-sealed document body parts,
pre-enciphered body parts, pre-enciphered document body parts, pre-sealed
document body parts, sealed document profiles and sealed profiles to Part 4.

The technical specifications of this Addendum are reflected at a number of places in Chaps. 3 and 4. An overview on the security aspects of
ODA documents is given in Sects. 3.5.

10.6   Addendum on Tiled Raster Graphics

The first version of Part 7 of the Standard did not contain the tiled raster
graphics encoding described in Chap. 7. The concept of tiling was added
to this Part of the Standard by the Addendum on Tiled Raster Graphics.

In particular, the attributes number of lines per tile, number of pels per
tile line, tiling offset and tile types were added and the value range of the
attribute type of coding in the context of raster graphics was extended to
provide for the tiled encoding.

The technical specifications of this Addendum are reflected in the description of the raster graphics content architecture in Chap. 7.

10.7   Expected Extensions

Work is currently underway on a number of extensions to the ODA Standard. Rather far advanced already are the Addendum on Streams which extends the functionality of the layout streams, and the Colour Addendum which adds a true colour model to the Standard. At present – in Spring 1991 – there are still several open issues which have to be resolved in the ISO and CCITT committees responsible for the ODA Standard and therefore these Addenda shall not be described here.

The Standard committees have also identified a number of further extensions to the ODA Standard such as

- an audio content architecture which provides for the inclusion of sound in ODA documents,
- extensions of the document structures to provide the creation and processing of so-called *hypermedia documents*, i.e., of documents whose structure is not a mere sequential order of the elements as is the case for printed documents,
- the inclusion of spreadsheets into ODA documents,
- a method for processing sophisticated tabular material,
- extensions to handle specialized character content such as mathematical or chemical formulae,
- providing an interface to used-definable applications, for instance, to retrieve data from data bases and include them in ODA documents, and
- a more general document processing model, including the semantic role of documents and their processing within an office environment.

At present, it cannot be described in much detail what these extensions will look like and, in particular, a time scale for their publication is hard to predict.

Index

The figures in the index indicate page numbers where a particular entry appears. A page number appears in bold when the entry is explained in more detail on this page.